Foundations in Signal Processing, Communications and Networking

Volume 23

Series Editors

Wolfgang Utschick, Technische Universität München, München, Germany
Holger Boche, Technische Universität München, München, Germany
Rudolf Mathar, RWTH Aachen University, ICT cubes, Aachen, Germany

This book series presents monographs about fundamental topics and trends in signal processing, communications and networking in the field of information technology. The main focus of the series is to contribute on mathematical foundations and methodologies for the understanding, modeling and optimization of technical systems driven by information technology. Besides classical topics of signal processing, communications and networking the scope of this series includes many topics which are comparably related to information technology, network theory, and control. All monographs will share a rigorous mathematical approach to the addressed topics and an information technology related context.

** Indexing: The books of this series are indexed in Scopus and zbMATH **

More information about this series at http://www.springer.com/series/7603

Riccardo Bassoli • Holger Boche • Christian Deppe
Roberto Ferrara • Frank H. P. Fitzek
Gisbert Janssen • Sajad Saeedinaeeni

Quantum Communication Networks

 Springer

Riccardo Bassoli
Technical University Dresden
Dresden, Germany

Christian Deppe
Technical University of Munich
Munich, Bayern, Germany

Frank H. P. Fitzek
TU Dresden
Dresden, Germany

Sajad Saeedinaeeni
Technical University Munich
Munich, Germany

Holger Boche
Technical University of Munich
Munich Center for Quantum
Science and Technology
Munich, Germany

CASA – Cyber Security in the
Age of Large-Scale Adversarie
Bochum, Germany

Roberto Ferrara
Technical University Munich
Munich, Germany

Gisbert Janssen
Technical University Munich
Munich, Germany

ISSN 1863-8538   ISSN 1863-8546   (electronic)
Foundations in Signal Processing, Communications and Networking
ISBN 978-3-030-62940-3   ISBN 978-3-030-62938-0   (eBook)
https://doi.org/10.1007/978-3-030-62938-0

This Springer imprint is published by the registered company Springer Nature Switzerland AG
The registered company address is: Gewerbestrasse 11, 6330 Cham, Switzerland

Preface

Taking a quick look at this book, a reader might think: *Oh, here comes another book on quantum computing, quantum information theory, and quantum communications!* This may be partially true. Quantum mechanics was born at the beginning of the last century and, over the decades, has obtained huge popularity in mathematics and physics. Additionally, quantum mechanics has been applied to computing and information theory for the past 40–50 years. Recently, it has been also applied to communications.

Thus, by exploring the available scientific literature, it is possible to find plenty of books on quantum mechanics, quantum computing, and quantum information theory and some on quantum communications. So, the question is: *Is this book needed in the panorama of scientific literature?* The answer is yes, and there are some important reasons to support this.

First, many of the books are not very recent (especially from a communication perspective), so they miss some important recent updates. Furthermore, most of them are monographs, which focus on specific areas of research in quantum theory and its applications.

Second, to the best of the authors' knowledge, no book has considered the recent perspectives that communication networks have been gradually acquiring. In fact, communication networks are currently undergoing a paradigm shift that adds computing and storage to the simple transportation ideas of our first communication networks. These *softwarized* solutions break new ground in reducing latency and increasing resilience but have an inherent problem due to the introduced computing latency and energy consumption. This problem can be solved by hybrid classical-quantum communication networks.

This book inherits the existing paradigm of computing-in-networks, and it uses this to describe future quantum communication networks (which will not only be the Quantum Internet). The book focuses on quantum computing, quantum information theory, quantum error correction, and system-level architecture as various bricks that will build future *compute-and-forward* quantum communication networks. The approach, which is used for the presentation of the theory of quantum communication networks, borrows some viewpoints from the ongoing work of the IETF Quantum Internet Research Group (qirg) (in which the authors are participating). However, this book also enhances and generalizes these views in order to leave the reader free to investigate and figure out new designs and solutions without architectural limits. This becomes especially important in a new field like quantum communication networks, where there are no existing standardized solutions yet.

Last but not least, this book addresses a topic in this field, which has never been presented before in books: the research problem of classically tested (via software simulations) quantum-mechanical systems. Because of this, at the end of the book, existing simulators of quantum communication networks are presented and their pros and cons are underlined. In this way, the reader will become aware of this important open issue when approaching research on quantum communication networks. Finally, some identified potential applications of quantum communication networks are also described. This also represents a practical viewpoint for the reader.

As authors who are experts in the fields of the research presented, we hope that the book conveys the importance of quantum-mechanical resources for the effective and efficient evolution of future communication networks. When we wrote this manuscript, we had in mind providing both physicists and engineers with a valuable reference for their research in quantum communication networks (and its subfields). Moreover, we planned the structure and the terminology to be both accurate and accessible, in order to become a helpful assistant for lectures in higher education and for training courses in the industry.

Dresden, Germany   Riccardo Bassoli
September 2020

Acknowledgments

We would like to thank Deutsche Telekom for supporting R. Bassoli and F. Fitzek over the last year and for their motivation to work on the topic of quantum communication networks.

We also thank the CeTI team for supporting R. Bassoli and F. Fitzek. CeTI is funded by the German Research Foundation (DFG, Deutsche Forschungsgemeinschaft) as part of Germany's Excellence Strategy—EXC 2050/1—Project ID 390696704—Cluster of Excellence "Centre for Tactile Internet with Human-in-the-Loop" (CeTI) of the Technische Universität Dresden.

We thank the German Research Foundation (DFG) within the Gottfried Wilhelm Leibniz Prize under grant BO 1734/20-1 for their support of H. Boche and C. Deppe.

Additionally, we thank the German Federal Ministry of Education and Research (BMBF) within the national initiative for "Q.Link.X—Quantum Link Extended" with the project "System Design for Secure Quantum Repeater Systems: Basic Protocols and Secure Implementation" under grant 16KIS0858 for their support of H. Boche, G. Janssen, and S. Saeedinaeeni and with the project "Quantum Information Theory and Communication Theory for Quantum Repeaters Beyond the Shannon Approach" under grant 16KIS0856 for their support of C. Deppe and R. Ferrara.

We also thank the German Federal Ministry of Education and Research (BMBF) within the national initiative for "Q.COM—Quantum Communication" with the project "Information Theory of the Quantum Repeater: Eavesdrop Secure Communication, Attacks and System Design" under grant 16KIS0118 for their support of H. Boche and G. Janssen and with the project "Eavesdrop Secure Communication via Quantum Repeaters When Using Different Resources" under grant 16KIS0117 for their support of C. Deppe.

Thanks also goes to the German Federal Ministry of Education and Research (BMBF) within the national initiative for "Post Shannon Communication (New-Com)" with the project "Basics, Simulation and Demonstration for New Communication Models" under grant 16KIS1003K for their support of H. Boche and with the project "Coding Theory and Coding Methods for New Communication Models" under grant 16KIS1005 for their support of C. Deppe and R. Ferrara.

Further, we thank the German Research Foundation (DFG) within Germany's Excellence Strategy EXC-2111—390814868 for their support of H. Boche and S. Saeedinaeeni.

In addition, we thank Werner Moorfeld for his continuous effort to bring different communities together. The many valuable discussions we had together formed the starting point for this book project.

Finally, we would like to thank E. Soeder for proofreading the book and making linguistic suggestions.

Contents

About the Authors

Riccardo Bassoli is a senior researcher at the Deutsche Telekom Chair of Communication Networks, Faculty of Electrical and Computer Engineering, Technische Universität Dresden (Germany). He received his B.Sc. and M.Sc. degrees in Telecommunications Engineering from the University of Modena and Reggio Emilia (Italy) in 2008 and 2010, respectively. Next, he received his Ph.D. degree from the 5G Innovation Centre at the University of Surrey (UK) in 2016. He was also a Marie Curie ESR at the Instituto de Telecomunicações (Portugal) and a visiting researcher at the Airbus Defence and Space (France). Between 2016 and 2019, he was a postdoctoral researcher at the University of Trento (Italy). He is an IEEE and ComSoc member. He is also a member of the Glue Technologies for Space Systems Technical Panel of IEEE AESS.

Holger Boche received the Dipl.-Ing. degree in electrical engineering, Graduate degree in mathematics, and the Dr.-Ing. degree in electrical engineering from the Technische Universität Dresden, Dresden, Germany, in 1990, 1992, and 1994, respectively, and the Dr. rer. nat. degree in pure mathematics from the Technische Universität Berlin, Berlin, Germany, in 1998. From 1994 to 1997, he did postgraduate studies at the Friedrich-Schiller Universität Jena. In 1997, he joined the Heinrich-Hertz-Institut (HHI) für Nachrichtentechnik Berlin, Berlin. From 2002 to 2010, he was a Full Professor in mobile communication networks at the Institute for Communications Systems, Technische Universität Berlin. In 2003, he became the Director of the Fraunhofer German-Sino Laboratory for Mobile Communications, Berlin, and in 2004, he became the Director of the Fraunhofer Institute for Telecommunications (HHI), Berlin, Germany. He is currently Full Professor at the Institute of Theoretical Information Technology, Technische Universitt Mnchen, which he joined in October 2010.

Christian Deppe received the Dipl.-Math. degree in mathematics from the Universität Bielefeld, Bielefeld, Germany, in 1996, and the Dr.-Math. degree in mathematics from the Universität Bielefeld, Bielefeld, Germany, in 1998. He was a research and teaching assistant at the Fakultät für Mathematik, Universität Bielefeld, from 1998 to 2010. From 2011 to 2013, he was project leader of the project "Sicherheit und Robustheit des Quanten-Repeaters" of the Federal Ministry of Education and Research at Fakultät für Mathematik, Universität Bielefeld. In 2014, he was supported by a DFG project at the Institute of Theoretical Information Technology, Technische Universität München. In 2015, he had a temporary professorship at the Fakultät für Mathematik und Informatik, Friedrich-Schiller Universität Jena. He is currently project leader of the project "Abhörsichere Kommunikation über Quanten-Repeater" of the Federal Ministry of Education and Research at the Fakultät für Mathematik, Universität Bielefeld. Since 2018, he is at the Department of Communications Engineering at the Technische Universität München.

Roberto Ferrara obtained his M.Sc. in physics at the Niels Bohr Institute of the University of Copenhagen and his Ph.D. in science at the Department of Mathematical Sciences of the University of Copenhagen. In his PhD dissertation "An Information-Theoretic Framework for Quantum Repeaters," he studied the limitations of distilling bipartite classical keys from quantum states when the two parties can only share entanglement with the aid of a third party, the quantum repeater, covering topics of entanglement measures, quantum operations, and quantum information theory. Since 2019, he is at the Department of Communications Engineering at the Technical University of Munich.

Frank H. P. Fitzek is a Professor and head of the Deutsche Telekom Chair of Communication Networks at the Technische Universität Dresden, coordinating the 5G Lab Germany. He is the spokesman of the DFG Cluster of Excellence CeTI. He received his diploma (Dipl.-Ing.) degree in electrical engineering from the University of Technology—Rheinisch-Westfälische Technische Hochschule (RWTH)—Aachen, Germany, in 1997 and his Ph.D. (Dr.-Ing.) in electrical engineering from the Technical University Berlin, Germany, in 2002 and became Adjunct Professor at the University of Ferrara, Italy, in the same year. In 2003, he joined the Aalborg University as Associate Professor and later became Professor. In 2005, he won the YRP award for the work on MIMO MDC and received the Young Elite Researcher Award of Denmark. He was selected to receive the NOKIA Champion Award several times in a row from 2007 to 2011. In 2008, he was awarded the Nokia Achievement Award for his work on cooperative networks. In 2011, he received the SAPERE AUDE research grant from the Danish government, and in 2012, he received the Vodafone Innovation prize. In 2015, he was awarded the honorary degree "Doctor Honoris Causa" by the Budapest University of Technology and Economics (BUTE).

Gisbert Janssen has been with the Institute for Theoretical Information Technology at the Technical University Munich as a researcher from 2010 to 2019. He received a physics diploma from the Berlin Technical University in 2010 and the Dr. rer. nat. degree from the Technical University Munich in 2016.

Sajad Saeedinaeeni received the B.S. and M.S. degrees in physics from the Royal Holloway University of London in 2010 and Leipzig University in 2015, respectively. He wrote his Master's thesis on Quantum Hypothesis Testing at the Max Planck Institute for Mathematics of Leipzig. He is currently working toward the Dr. rer. nat degree in physics at the Institute of Theoretical Information Technology of the Technische Universität München (TUM) under the supervision of Holger Boche.

Chapter 1
Introduction

The rise of new fundamental theories in physics has always opened the door for subsequent advancement in practical physics and theoretical engineering. For example, the discovery of electromagnetism in the nineteenth century resulted in the study, design, and development of telecommunications and computing during the following twentieth century. The fundamental theory of the last century is quantum mechanics, which has had significant scientific and philosophical impact by changing the way we look at and interpret the universe. In fact, in the mid twentieth century, preliminary formulations of quantum mechanics became more and more mature by subsequently revealing, in the late twentieth century, the potential engineering perspectives of quantum phenomena such as photonics, computing, and cryptography. Furthermore, in the last two decades, quantum-mechanical resources became the main suppliers for an infrastructural evolution of existing communication networks, in order to make them capable of addressing the existing challenges in computing and telecommunications.

1.1 The Evolution of Classical Communication Networks

The first commercial worldwide communication networks emerged with the well-known telephone services, which initially required direct links between all communication partners. Next, scalability was improved by introducing localized central switching to reuse telephone cables more efficiently. The concept of hierarchical switching was introduced into telephone networks with the rise of circuit switching. Especially in circuit-switched networks, communication pairs always use dedicated and exclusive physical resources. Though realized by utilizing several hops across connecting equipment, the resulting implementation logically appeared as a single dedicated virtual cable between communication partners.

R. Bassoli et al., *Quantum Communication Networks*, Foundations in Signal Processing, Communications and Networking 23, https://doi.org/10.1007/978-3-030-62938-0_1

Next, the era of packet switching started. This communication paradigm enabled the breaking up of long messages into smaller ones and realized the concepts of efficient time-shared resource utilization. In 1974, a unifying protocol suite for heterogeneous communications was needed on top of various packet-switched networks, to enable interoperability. This specific suite was composed of the transport control protocol (TCP) and the internet protocol (IP). Jointly, these two layers of protocols enabled a process-oriented and reliable communication end to end, across different packet-switched networks. It was this addition to packet switching that enabled the seamless interconnection of individual networks to the Internet. With the evolution of the Internet, the idea of packets and packet switching following the *store-and-forward* policy was fully adopted.

However, future networks will have a completely different nature from both existing wireless and wired perspectives. That is why, nowadays, standardization is undertaken by 3GPP (wireless part) together with IETF (wired part). In fact, the current objective is to realize an *ecosystem* (or *pan infrastructure*), capable of interconnecting highly heterogeneous networks, achieving concurrently demanding requirements and supporting several different verticals. The main enabler to achieve this scope is virtualization: the deployment of software-defined networking (SDN) [NMN+14] and network function virtualization (NFV) [MSG+16] at all levels. The main characteristic of network virtualization is the software-based implementation of functions, protocols, and operations, running on general purpose hardware.

The so-called process of *softwarization* of the network has produced the fertile ground for efficient and effective novel paradigms such as cloud and edge/fog computing, unique and flexible/reconfigurable SDN–NFV architecture, and end-to-end network slicing [RBSG16, LSC+17] (with complete isolation among slices and services). In fact, significant research effort has been focused on the design of a common/unique SDN–NFV system. Furthermore, extreme flexibility is targeted by the new paradigm of the wireless network operating system (WNOS), which will completely abstract network entities by providing a programmable protocol stack (PPS), rather than only network functions and routing.

Such a complex and heterogeneous system will require future networks to rely less on human intervention and more on machine learning/cognition for network management. In fact, an increasing research trend focuses on the deployment of cognition to make network management autonomous. The research community has started studying self-organized networks (SONs) in parallel to virtualization by extensively applying machine learning and cognitive algorithms toward self-healing and self-management.

The vision of future networks implies the realization of cloud and edge computing in a more and more distributed manner, in order to respond to different legal and technical requirements, while increasing resilience in data storage and computing. Computing will become an intrinsic characteristic of future networks. The placement of computing at a given data center (e.g., big, micro, femto, etc.) will have impact on security, resilience, capacity, and latency of a given end-to-end communication. Furthermore, the distributed nature of the future computing paradigm may also be extended to end users. Because of the prominent role, computing will have in future generation networks, especially when performed in a distributed manner, the intrinsic nature of the network will experience a radical transformation of the paradigm: from solely conveying information between two places using *store and forward* to where information is also processed within the communication network, using *compute and forward* [FGS20].

The characteristics of future networks briefly described above predict the rise of a very high demand for storage capacity and computing infrastructure. This will also significantly increase energy consumption in communication networks. Moreover, the realization of intelligent and adaptable networks will require a huge number of resources for secure data mining/processing and distributed computing for decision-making. Intelligent analysis of Big Data will continuously need network performance and network infrastructure awareness for prediction of future network states. Additionally, the synchronization of highly distributed computing entities will also affect requirements in terms of capacity, latency, and reliability.

Regarding performances of future communication networks, some of the desired key performance indicators (KPIs) [ARS16] are: 1 ms round trip latency, billions of connected devices, perceived availability of 99.999%, and reduction in energy usage by almost 90%. These KPI goals will allow 5G and beyond networks to support several possible end-to-end communication paradigms/services. Such services, also named *verticals*, are usually grouped into three main categories: Extreme Mobile Broadband (xMBB), ultra-reliable Machine-Type Communications (uMTCs)—also called Ultra-Reliable Low-Latency Communications (URLLCs)—and massive Machine-Type Communications (mMTCs).

The concurrent satisfaction of these requisites, via algorithms and solutions described above, is limited to the existing domain of communications based on classical physics. These imply an enormous effort and complexity but do not ensure a successful solution. The following underlines why the intrinsic drawbacks of future generation networks will limit their capabilities.

The process of *softwarization* in future networks must attain a high flexibility toward a reduction in operational expenditure (OpEx) and capital expenditure (CapEx). Software and virtualization can be viewed as the DNA of 5G and beyond networks' infrastructure. However, software abstraction also introduces additional packet processing delays. In fact, it increases data transfer and computation demands so that latency becomes relatively high. Even if the existing virtual switches are quite fast, virtual machines (VMs), specifically the packet IO and processing operations inside the VM, are slow. Two software bridges (or virtual switches) connect virtual network interfaces (vNICs) with the physical network interface (pNIC). Next, the integration bridge connects all VMs running on the same physical node.

A proposed centralized approach to virtualization gives more than 2 ms latency inside the virtual communication environment for a single VM running the elementary forwarding function [XGU+19], which becomes unacceptable in the context of URLLCs. Some solutions have been designed to significantly reduce the latency of virtualization [XGU+19], but more and more software is getting into network layers so that latency due to network virtualization is expected to constantly increase. Furthermore, some components of latency are proportional to the available transmission bitrate and to the amount of data. This can be addressed by scaling up the transmission capacities and compression, but processing delays with their various constant delays still affect latency with their contribution.

When employing network virtualization, further important aspects also have to be taken into account: reliability and resilience (i.e., keeping an acceptable level of service in case of faults). In fact, software-based network functions and applications are more prone to failures than hardware-based ones because they have more points of faultiness. This makes it harder to match the desired network flexibility with future KPIs of network reliability and service availability [LJK$^+$16, GB18]. Moreover, the realization of cloud and edge computing in a more and more distributed manner also includes the impact of data storage and computing at data centers. In fact, the placement of computing at a given data center (e.g., big, micro, femto, etc.) has impact on resilience, capacity, and latency of a given end-to-end communication. Furthermore, the distributed nature of the future computing paradigm could also be extended to the user end device, adding more points of delay and failure.

While making software the pillar of future communication networks improves some performance metrics as mentioned above, this also opens various new security threats. A major security issue of SDN–NFV-based systems is the threat due to scalability. Controllers and hypervisors can easily become bottlenecks because of the amount of control traffic they have to manage. Control plane saturation opens the door to various Denial-of-Service (DoS) and distributed DoS (DDoS) attacks. Regarding network management and orchestration, authenticating applications becomes fundamental. It is necessary to establish a trust relationship between the control plane and applications. Next, in SDN, effective access control and account-ability mechanisms are still an open challenge. Medium access control, SDN can be attacked by a malicious host, sending packets with falsified source Medium access control addresses to poison the Medium access control table of a switch. Many other security threats due to network *softwarization* and network function computing can be found in [ANYG15, DCA$^+$17, FTKS19, SNS16, AvW18, PHS$^+$18].

The switch of the paradigm from *store and forward* to *compute and forward* will massively increase the deployed computing resources in the network and the size and/or number of data centers. This implies a significant rise in energy consumption because of virtualization and *softwarization* [DWF16, JHA$^+$16, JITT16, BIK$^+$16, BGADR19]. The consumption of a big data center is about 25.000 households, and the energy costs of powering a data center double every 5 years (and this rate will grow due to massive computing deployment) [DWF16]. Such energy needs also imply environmental problems (e.g., in 2005, the total data center power consumption was 1% of the total US power consumption, creating the same quantity of emissions as a mid-sized country [DWF16]). The energy required by computing can be divided into two main categories: energy used by network/computing equipment (e.g., servers, networks, storage, etc.) and energy used by infrastructure facilities (such as cooling, air conditioning, etc.).

Moreover, the realization of intelligent adaptable networks will require huge amounts of resources for secure data mining/processing and distributed computing for decision-making. Big data intelligent analysis will continuously need network performance and network infrastructure awareness for prediction of future network states. Additionally, the synchronization of highly distributed computing entities will also affect requirements in terms of capacity, latency, and reliability.

Classical future generation networks also present a *communication complexity problem*. Communication complexity is the amount of information (in terms of bits) that spatially separate computing devices need to exchange in order to successfully perform a computational task. Virtualization and computing paradigms such as mobile edge computing (MEC) and distributed computing for network functions will rely on distributed devices solving network related computing problems. However, some of these distributed computing problems were demonstrated not to be solvable via distributed computing based on classical networks. Alongside solvable problems, the amount of communication bits used for communication among distributed computing devices was shown to have a strong impact on the links of the network (techniques for compression and encoding are not able to limit this growth effectively) [BCMdW10]. Such a problem of complexity logically recalls the previously mentioned latency one because delay is also proportional to the available transmission bitrate.

Finally, the integration of multiple services today is realized by higher-layer policies that allocate different services on different logical channels. The security issue is usually addressed by applying cryptographic techniques at higher levels. In general, this is quite inefficient, and there is a trend to merge multiple coexisting services efficiently so that they work on the same wireless resources. This is referred to as physical layer service integration [SB14a] and has the potential to significantly increase the spectral efficiency for next-generation networks. It is expected that different applications (e.g., secure message transmission, broadcasting of common messages, and message transmission) will all be implemented by *physical layer service integration*. References [HW10a, DS05] have been the first papers for quantum systems that offer a larger variety of services.

1.2   Toward Quantum Communication Networks

The previous section presented the path toward the current new paradigm promoted by future 5G and beyond networks. Communication networks have evolved from circuit-switched to packet-switched networks. In both cases, the end-to-end design has dominated, allowing for the transparent transportation of bits as given in Fig. 1.1. In such a system, latency is impacted by the propagation delay of the communication link.

Next, to reduce propagation delay, computing and storage have been added to communication nodes, in order to allow proximity-based computation (scheme in the second line of Fig. 1.1). By means of the virtualization paradigms mentioned above, the propagation latency has been significantly reduced. The MEC has been especially fundamental for such a purpose.

However, even if virtualization has contributed to reduce communication latency due to propagation, it has actually increased latency in every computing virtual

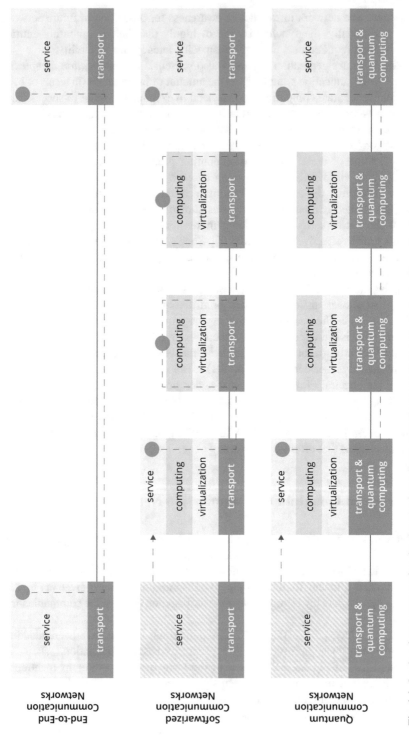

Fig. 1.1 Logic scheme representing the change of paradigms according to evolution of communication networks

machine [XGU⁺19] involved. The reason is that all processed bits of information have to pass through several software-based virtualization layers (see Fig. 1.1, scheme in the middle). Thus, not only existing but also future solutions for *softwarized* networks can only support applications with a small number of virtual machines. Such limitation is necessary in order to not too rapidly increase networks' computing complexity.

All the aforementioned disadvantages expressed here and in the previous section, due to intrinsic classical limitations of both virtualization and computing, can only be resolved via the development of intrinsically different networks based on quantum mechanics. In this way, classical network computing is transferred to quantum-mechanical network computing: the change of the paradigm moves virtual network/computing resources based on software to new physical quantum resources relying on entanglement, which does not reveal itself in classical systems.

Just as classical networks were initially built on top of the existing legacy telephone infrastructure, current quantum-mechanical networks are going to be built in a hybrid fashion, on top of the existing classical network infrastructures.

By employing distributed quantum computing instead of classical computing, a unique virtual quantum machine consisting of a high number of entangled qubits scaling with the number of interconnected devices can achieve an exponential speed-up of a network's computing capabilities with just a linear increase in physical resources. In such a way, the limitations imposed by classical paradigms on virtualization and *softwarization* can be exceeded by exploiting quantum-physical parallelism based on the concepts of quantum superposition, entanglement, and quantum measurement (see the bottom of Fig. 1.1).

Existing approaches, using quantum mechanics in networks to provide security, mainly carry out protocols for key exchanges between individual nodes. In many cases, quantum key distribution (QKD) is used from one node to another. On the other hand, there are limited analyses focused on the complexity and capacity of these processes. Therefore, the current research and investments are principally pivoted on military and governmental communications, while civil users are strongly concerned with speed and complexity (which significantly affect latency and quality of the communications).

The potential of quantum mechanics for communication systems was recognized early on. Holevo's work [Hol73] in 1973 considers classical communication via quantum channels. In 1984, Bennett and Brassard developed the first quantum cryptography protocol in [BB84]. There, demonstrably secure communication theory was mentioned and considered. At the beginning of 1990, new quantum protocols for the solution of quantum communication tasks over quantum channels were developed. This included, for example, quantum teleportation. This is a process in which quantum information can be transmitted from one location to another, with the help of classical communication and previously shared quantum entanglement between the sending and receiving location. The seminal paper on this was published by Bennett, Brassard, Crépeau, Jozsa, Peres, and Wootters in 1993 [BBC⁺93].

Furthermore, many classic coding methods and information theory approaches that were shown during this time do not apply quantum-mechanically, due to Heisenberg's uncertainty principle. A prominent example is the no-cloning theorem [WZ82]. After 2000, several exact information-theoretical capacity results could be proven. These results include, for example, private transmission [Dev05], distillation of secret key and entanglement from quantum states [DW05], and quantum wiretap channels [CWY04]. All experimental systems today essentially provide key generation in point-to-point communication.

The idea behind quantum communication networks [SR00, Kim08, WEH18] is that nodes of the network may be considered as distributed parts of the same physical system. The reason why quantum communication networks can outperform the classical Internet with relatively modest resources is because it comes from inherent properties of quantum-mechanical systems. In particular, entanglement is used as a resource for communication.

Entanglement (*Verschränkung*) is the main pillar for effective distributed quantum computing, teleportation, and superdense coding (to achieve efficiently higher throughput). Moreover, entanglement can also be interpreted from an information theoretic perspective. Quantum information theory is an important aspect of both quantum computing and quantum communications. Additional quantum-mechanical aspects are also the impossibility of copying information (no-cloning theorem) and qubits (superposition of possible states).

Quantum-mechanical-based computing and networking seem to be fundamental paradigms that will efficiently and effectively solve the existing problems that research and industry have been encountering in the design and development of future networks. In fact, technologies based on quantum mechanics (mainly exploiting entanglement) have been demonstrated to be effective for distributed systems while also guaranteeing intrinsic security of communication and computing in the cloud. Moreover, the realization of small quantum-based distributed and interconnected computing entities is the quantum more effective version of the current MEC paradigm. As an example, scientists can leverage distributed quantum computing to solve highly complex scientific computation problems such as the analysis of chemical interactions for medical drug development. This is not efficiently and effectively possible via classical networks and computing.

Regarding the current effort toward intelligent networks based on data mining and distributed data computing, quantum machine learning algorithms running on distributed quantum computing nodes can be secure, effective, and efficient (in terms of usage of network resources and energy).

Furthermore, quantum communications have provided significant advantages in communications, where multiple sources transmit information to a single receiver via a unique communication channel [Nö19, LALS20]. In particular, the comparison between classical and quantum multiple access channels (MAC), where two sources transmit to a destination, proved that quantum communications exceeded the success probability of classical ones because of the exploitation of entanglement.

On the other hand, apart from exceeding the capabilities of current classical networks, quantum networks can also be a fundamental means for other aspects of communication networks. Complex quantum networking processes can be realized at the quantum level, between atomic ensembles. These special quantum networks can be used to generate arbitrary connections inside the network, achieving truly random behavior. Then, with the help of these quantum random networks, complex real-world problems can be analyzed and modeled. Thus, random quantum networks could help to model real-life communication, in physical or biological systems, and could be extended to model the behavior of global communication networks (the classical existing Internet). By using quantum communications and the intrinsic stochastic nature of quantum systems, the still-undiscovered connections in real-life communications can be also studied more effectively [IG12].

Finally, quantum communication networks can solve the *communication complexity problem* mentioned in the previous section. Let us consider different network nodes. First, they own data and are forbidden from having any communication among themselves. Their aim is to compute a function of their data to produce output bits. Such a computational task (i.e., the computation of a relation) is nontrivial since both the owned and output bits are distributed among the parties; thus, each node only has the value of one of their subsets. As mentioned above, classical communications must have transmissions among the parties in order to succeed. While entanglement cannot be used to replace communications, it can be employed to significantly reduce the communication load [BCMdW10].

The parties can communicate to each other via bits or qubits, *classically* sharing randomness, or alternatively, sharing entangled quantum states, if quantum communications are considered (this concept is further clarified in Sect. 2.5). What is the minimum number of communication bits or qubits that are needed for computing? Entanglement cannot be used for signaling but can reduce the communication needed for distributed computing. Depending on the computing problem, communication complexity can be reduced to $n/2$ (due to superdense coding), or even more, such as to $\sqrt{n}$ and $\frac{n^{1/4}}{\log n}$, given n the number of bits that should have been transmitted in the case of classical communications [BCMdW10].

1.3 Structure of the Book

This book aims at equipping the readers with the fundamental background and guidelines to investigate and to design quantum communications systems with an eye toward future generation networks. In order to do this, the book explains in a comprehensible manner, fundamental concepts of quantum mechanics, quantum information theory, quantum computing, and quantum communication networks as a whole, considering the perspective of both physicists and engineers.

The book tries to balance between conveying content in an accessible manner and being accurate and precise. In this way, a reader will obtain fundamental knowledge and deeply understand the research in quantum communication networks (QCNs).

Figure 1.2 depicts a Venn logic diagram of the subject *quantum communication networks*. This research field is the union of sub-areas such as quantum computing and quantum communications. These two macro-subjects intersect and rely on

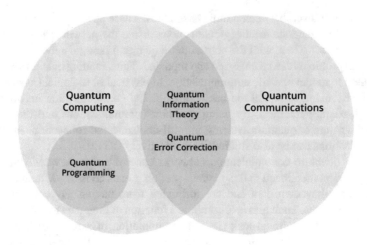

Fig. 1.2 Venn logic diagram representing the main research sub-areas (and their relations) in the research field of quantum communication networks

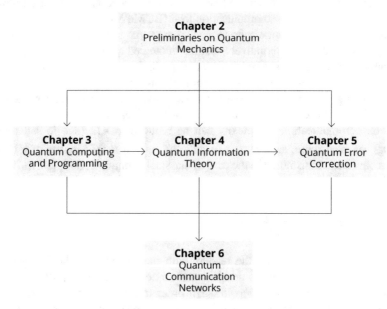

Fig. 1.3 Organization of the book and logic interconnections among chapters with suggested directions for reading

quantum information theory and quantum error correction. Finally, the subject area quantum computing includes the problem of how to program quantum hardware.

Next, Fig. 1.3 maps the Venn logic diagram into the various book chapters, also displaying the suggested direction of reading. Chapter 2 is important for the fundamental concepts, which are needed to understand all the results and ideas in quantum communication networks. Subsequently, Chapter 3 provides the basics on quantum computing and quantum circuits, which are the pillars for describing quantum algorithms, various concepts on quantum information theory, and building blocks of quantum communication networks. Chapter 4 explores the fundamental basics and achievements related to quantum information and quantum channels. Next, Chapter 5 analyzes the most important error correction techniques for quantum communication networks. Finally, Chapter 6 addresses the design of quantum communication networks by showing all the building blocks required to set up a quantum communication. Furthermore, it also describes the most well-known simulation tools internationally available, by highlighting the fundamental issues behind the question *how to classically test and evaluate (via software running on classical machines) the performance of quantum communication networks.*

Chapter 2
Fundamental Background

Traditionally, telecommunications theory has relied on classical physics and rules of classical systems; however, telecommunications actually obeys the more general postulates and laws of quantum mechanics. In order to take full advantage of telecommunication, we thus need to take into account and encompass the ideas of quantum communication and quantum networks, which intrinsically change its nature. The goal of this book is to gain access to all the features of telecommunications. In this context, quantum mechanics is taken simply as a tool to achieve the correct predictions, without making any claim on the correctness of the foundations of quantum mechanics, which are presented in Sect. 2.6 for completeness. This chapter introduces the notions, tools, and notations used in quantum theory. The differences between the tools provided for classical and quantum theory are summarized in Table 2.1.

2.1 Preliminaries on Quantum Mechanics

This section will introduce the axioms of quantum mechanics necessary for the design of quantum networks. We will see how to describe the single and composite quantum systems analogue of deterministic classical systems, while the modeling of the noise and thus of general quantum information and statistics is left for the next section. More details about the important concepts of quantum mechanics presented in this section can be found in [Bel06, AFP09, NC10, BA12].

© The Editor(s) (if applicable) and The Author(s), under exclusive license to
Springer Nature Switzerland AG 2021
R. Bassoli et al., *Quantum Communication Networks*, Foundations in Signal
Processing, Communications and Networking 23,
https://doi.org/10.1007/978-3-030-62938-0_2

Table 2.1 Preview of the changes from the classical to the quantum theory of finite discrete variables

Object	Classical system	Quantum system
States	Elements of a finite set	Unit vectors in finite dimensions
State changes	Functions	Unitary matrices
Random states	Probability distributions	Positive matrices with trace one
Channels (linear maps between random states)	Stochastic matrices	CPTP maps
Physical quantities	Random variables	Hermitian matrices

2.1.1 Postulates of Quantum Mechanics

The technological applications of quantum mechanics such as quantum information, quantum communications, and quantum computing became possible, as for any theory, because of rigorous formalism and mathematical description. The current formulation of quantum mechanics comes from the work of von Neumann [vH58], who formalized its fundamental postulates.

In 1925, the first consistent mathematical formulation of quantum mechanics was the so-called *matrix mechanics*, which was jointly proposed by Werner Heisenberg, Max Born, and Pascual Jordan. As its name suggests, this theory describes the physical properties of particles as matrices, the observables evolving in time. Side by side, in 1926, Erwin Schrödinger formulated a partial differential equation describing the time evolution of the state of closed quantum systems, theorizing the concept that subatomic particles owned properties of electromagnetic waves and classical particles concurrently: in particular, the resulting wave function defined the state of the quantum system at each spatial position and time. Today, in the modern formulation, the two are known as the Heisenberg and Schrödinger representations, which differ only on whether the evolution is applied to the state or to the observable.

Next, in 1930 and 1932 respectively, Paul Dirac and John von Neumann demonstrated that matrix mechanics and Schrödinger's equation were two different approaches to the same theory. They proposed the so-called *Dirac–von Neumann axioms* by providing a theoretical description of quantum mechanics in terms of operators on a Hilbert space. Figure 2.1 summarizes the main steps towards current mathematical formalism.

The theory of quantum mechanics is now based on some fundamental postulates derived from experimental results.

Postulate 1 claims that states of quantum systems can be described via vectors in a complex Hilbert space (or density matrices as we will see later). A linear decomposition into vector states is also called a *superposition* of vector states. In particular, given a classical system, its quantization[1] will assign a vector to each

---

[1]Not to be confused with the classical quantization of a signal, which is a discretization of the values of a continuous signal, usually performed after sampling.

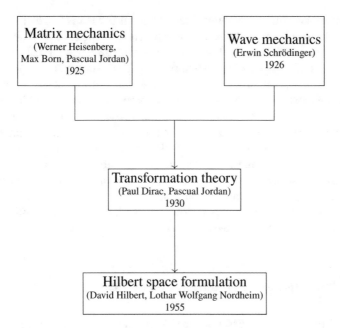

Fig. 2.1 Research path towards achieving the latest mathematical formulation of quantum mechanics

classical state and the space of quantum states will be a Hilbert space in this vector space.

Postulate 2 states that the evolution in time of a closed quantum system follows Schrödinger's equation, which implies that the evolution of the quantum state follows a unitary evolution.

Next, von Neumann additionally identified measurements as a different type of evolution which cannot be described by Schrödinger's equation. In fact, a measurement implies coupling (entanglement) between the quantum system under measurement and the equipment/environment for measurement, thus becoming an open system.

Postulate 3 claims that a physical observable is a Hermitian operator. Measuring an observable yields one of its possible eigenvalues, with the probability given by the magnitude of the normalized state vector onto the eigenspace of the outcome. Contrarily to Postulate 2, this causes a loss of natural quantum coherence and the evolution of the quantum system generally implies an irreversible loss of information and the destruction of the superpositions across the eigenspaces.

Finally, *Postulate 4* refers to the description of states of composite quantum systems. The Hilbert space of a composite system is represented by the tensor/Kronecker product of the Hilbert spaces of single subsystems. Similarly, the space of composite observables is the space of Hermitian operators of the tensor product Hilbert space. This postulate is the one that allows the description of entanglement and entangled observables because it does not state that the possible state and observables are only the tensor products of single states and observables.

In the rest of the section these postulates will be made formal. In classical physics, dynamical systems are characterized by functions, which describe the time-dependence of a point in space. Furthermore, the possible states a system can assume are represented in the so-called phase space, where a unique point represents a state to which specific values of observables are assigned. In particular, classical physics considers observables as real-valued functions over the phase space. On the other hand, quantum mechanics requires the states of a dynamical system and its observables to obey the linearity of an underlying Hilbert space, while at the same time describing classical theory as a special case. Thus the new quantities introduced by the theory of quantum-mechanical systems must have a different nature. The Dirac–von Neumann formulation provides a theoretical description justified by full agreement and consistency between theory and final experimental results.

2.1.2 Formulation of Quantum Mechanics

The new formulation must describe the superposition and vector nature of quantum states and therefore we have to employ vectors spaces. However, ordinary finite-dimensional vectors are not enough to describe general systems in quantum mechanics, and even though we will only consider finite-dimensional quantum systems, in general, infinite-dimensional vector spaces must be considered. This is already the case, for example, in the quantization of the harmonic oscillator, and thus of the quantization of electromagnetic modes.

In order to manipulate all vectors states in a unified manner, Dirac developed a suitable notation, the so-called *Dirac's bra-ket notation* or *bra-ket notation*. In this notation, vectors are denoted by $|\psi\rangle \in V$, also called *kets*. A basis decomposition in the finite dimension then looks like

$$|\psi\rangle = \sum_{i=1}^{n} c_i |u_i\rangle \equiv \sum_{i=1}^{n} c_i \mathbf{u}_i \tag{2.1}$$

where coefficients $\{c_i\}$ are complex numbers, which multiply a basis $\{|u_i\rangle\}$. In quantum theory the coefficients c_i are also called *amplitudes*, due to their role in expectation values, as we will see below. Elements of the dual vector space $V^\dagger$ and the space scalar linear operators V are denoted by $\langle\psi|$, also called *bras*. V is also the space of scalar linear operators on $V^\dagger$ and the two spaces are dual to each other. In finite dimensions, V and $V^\dagger$ have the same structure and dimension, namely there exists an invertible mapping from vectors to dual vectors. Then it is possible to write

$$|\psi\rangle \in V \Leftrightarrow \langle\psi| \in V^\dagger \tag{2.2}$$

$$|\psi\rangle^\dagger = \langle\psi|. \tag{2.3}$$

Fig. 2.2 Representation of
the scalar product between
bras and kets

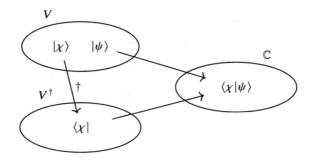

In complex vector space, with the usual inner product, this mapping is the Hermitian conjugate and we can write $(c|\psi\rangle)^\dagger = c^* \langle\psi|$, where c^* is the complex conjugate of the complex number c. So by definition, bras and kets are complex quantities, since they can be multiplied by complex numbers. There is no especially defined notation for complex conjugation and transposition, which reflects the fact that these are not physical operations on quantum states. In particular, extracting the real or complex part of a state vector has no physical meaning. Bras and kets are elements of dual vector spaces; however, the spaces are different and thus bras and kets cannot be summed together.

In spaces with a scalar/inner product (a complex number in complex vector spaces), the inner product between two vectors can be seen as simply the action of one of the dual vectors onto the remaining vector, as shown in Fig. 2.2 in Dirac's notation. In Dirac's notation, the *inner product* (also called the *braket*) between two vectors $|\chi\rangle, |\psi\rangle \in \mathbb{C}$ is denoted by

$$\langle\chi|\psi\rangle \in \mathbb{C} \tag{2.4}$$

which makes it seamless in interpreting it as an inner product, as the action of a bra on a ket, or as the action of a ket on a bra. More explicitly, it is important to note that $\langle\chi| = |\chi\rangle^\dagger \in V^\dagger$. The inner product can be viewed as mapping ket $|\chi\rangle \in V$ onto its bra image $\langle\chi| \in V^\dagger$ using $\langle\chi| = |\chi\rangle^\dagger$, and then, since $\langle\chi|$ is a function on V, computing its action on $|\psi\rangle$ as $\langle\chi|\psi\rangle = \langle\chi|(|\psi\rangle)$. In complex spaces it holds that $\langle\chi|\psi\rangle = \langle\psi|\chi\rangle^*$. Two vectors are orthogonal if their inner product is zero and in Dirac's notation for two kets $|\chi\rangle, |\psi\rangle$, the condition is thus written as

$$\langle\chi|\psi\rangle = 0. \tag{2.5}$$

Two states of a quantum system are said to be *orthogonal* if the corresponding vectors are orthogonal. Similarly, changing the notation does not change the definition of the norm, which is still defined as

$$\| |\psi\rangle \|^2 := \langle\psi|\psi\rangle \tag{2.6}$$

$\lvert \cdot \rangle + \lvert \cdot \rangle$	sum of kets $\longrightarrow$ complex vector
$\langle \cdot \rvert + \langle \cdot \rvert$	sum of bras $\longrightarrow$ complex vector
$\langle \cdot \rvert \cdot \rangle$	scalar product (braket) $\longrightarrow$ complex number
$\lvert \cdot \rangle \langle \cdot \rvert$	outer product $\longrightarrow$ complex matrix

Fig. 2.3 Allowed operations with bras and kets and their respective results

which is real and non-negative. Figure 2.3 lists the operations allowed with bras and kets and the respective results.

The complete algebraic scheme involves three main quantities: bras, kets, and linear operators. The action of a linear operator A on a vector $\lvert \psi \rangle$ in Dirac's notation is simply $A \lvert \psi \rangle$, corresponding to the right multiplication $A \cdot \mathbf{v}$ in the usual vector notation. If $\lvert \chi \rangle$ is a vector in the output space of A, the inner product between $\lvert \chi \rangle$ and $A \lvert \psi \rangle$ is simply[2]

$$\langle \chi \rvert A \lvert \psi \rangle . \tag{2.7}$$

Dirac's notation seamlessly shows that A is also an operator between dual vector spaces mapping $\langle \chi \rvert$ to $\langle \chi \rvert A$, as much as it is an operator between vector spaces. If A is a linear operator from vector space V to vector space W, then it is also a linear operator from $W^\dagger$ to $V^\dagger$. This is not in contradiction with complex linear operators being sesquilinear maps, because a complex matrix A is conjugate-linear in the first argument of $A(w, v) = (w, A \cdot v)$ only as a map from $W \times V \to \mathbb{C}$.

The definition of the adjoint (or Hermitian conjugate if the inner product is the complex inner product) $A^\dagger$ also becomes visually more expressive, being defined as the unique operator from W to V, such that $A^\dagger \lvert \chi \rangle = (\langle \chi \rvert A)^\dagger$.[3] This can be viewed as a special case of $(AB)^\dagger = B^\dagger A^\dagger$, as indeed both kets and bras are special cases of linear operators where one of the vectors spaces is $\mathbb{C}$, namely they are, respectively, linear operators from $\mathbb{C} \to V$ (via scalar multiplication) and from $V \to \mathbb{C}$ (via the inner product).

In order to state the rest of the quantum postulates, we must have the ability to choose an orthonormal basis compatible with the inner product, in the sense that the inner product can be written as a convergent sum over the coefficients of

---

[2] A *linear operator* between vector spaces V and W is defined to be any function $A : V \to W$, which is linear in its inputs, such that $A \left(\sum_i a_i \lvert v_i \rangle \right) = \sum_i a_i A(\lvert v_i \rangle)$, where $A(\lvert v \rangle)$ can also be written as $A \lvert v \rangle$. When a linear operator is defined on a single vector space, it means $A : V \to V$. Two important examples are the identity operator $I_V \lvert v \rangle \equiv \lvert v \rangle$ and the zero operator $0 \lvert v \rangle \equiv 0$. An $m \times n$ complex matrix A is a linear operator $A : \mathbb{C}^n \to \mathbb{C}^m$, under the matrix multiplication of the matrix A by a vector in $\mathbb{C}^n$.

[3] In contrast, in standard notation the adjoint is the unique linear operator such that the unique linear operator $A^\dagger$ on V is $(v, Aw) = (A^\dagger v, w)$. Hermitian conjugation in finite dimensions and matrix representation is reduced to complex conjugation and transposition.

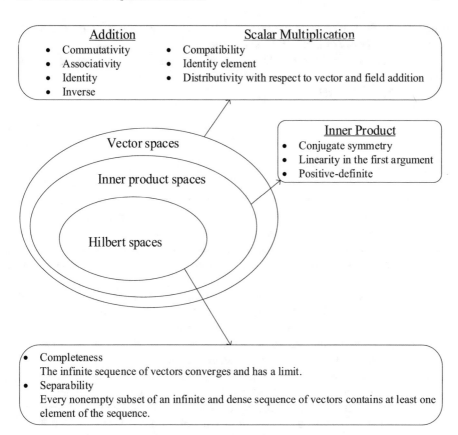

Fig. 2.4 Classification of different kinds of vector spaces according to their properties

the basis. Such an inner-product space is called a *Hilbert space*. Figure 2.4 clearly depicts this relation between vector spaces, inner-product spaces, and Hilbert spaces according to their properties. In finite dimensions these properties are automatically satisfied, and any inner-product space is a Hilbert space. If the dimension is finite, the properties can be proved by defining the inner product, otherwise, if the Hilbert space is infinite-dimensional, they should be considered as axioms. The mathematical description of quantum systems often requires the structure of infinite-dimensional inner-product spaces. In such cases, proper convergence and limiting procedures should be taken into account.

Bras and kets (or their directions) in a Hilbert space $\mathcal{H}$ physically represent states of a physical system, usually directly identified with its Hilbert space $\mathcal{H}$, while Hermitian operators ($A = A^\dagger$) on $\mathcal{H}$ correspond to physical observables. As previously mentioned, dynamical variables or observables are quantities, such as the coordinates and components of velocity, momentum and angular momentum of particles, and functions of these quantities. These are fundamental variables in classical mechanics, which should also occur in quantum mechanics. However, the main intrinsic difference between classical and quantum mechanics is that the commutative axiom of multiplication does not hold.

The heuristic expectation value of an observable A on a certain state $|\psi\rangle$, estimated through repeated experiment, will converge to the physical *expectation value* of A, which is defined as

$$\langle A \rangle := \frac{1}{\langle \psi | \psi \rangle} \langle \psi | A | \psi \rangle . \tag{2.8}$$

The reason for the axiomatic normalization is that to maintain the interpretation as an expectation value, the expectation value of a constant observable $c \equiv c\mathbb{1}$ should be only the constant c itself. The result is that quantum vector states are equivalent up to normalization and to a phase factor e^{iy} where $y \in \mathbb{R}$, because they produce the same expectation values on all observables. What determines the quantum state is the direction, or ray, in the complex vector space. In order to remove this freedom, all quantum states are assumed to be normalized to one, and can be assumed to be orthogonal rank-one projectors, like $|\psi\rangle\langle\psi|$.

Hermitian operators are the analogue of random variables—however some precautions are needed. For two random variables X and Y, the common implicit assumption is that they are described by a joint probability distribution (the state), rather than being different variables on the same distribution, like X and $f(X)$ for some function f. All observables on $\mathcal{H}$ share the same state, in particular for any analytic function f with domain containing the spectrum of A. The operator $f(A)$ is well defined[4] and the two are the analogue of the random variables X and $f(X)$. In particular, a vector $\mathbf{A} = (A_1, \ldots, A_n)$ of observables is the analogue of a random vector and not the analogue of a vector of random variables. However, if f is not injective, measuring A and then computing f on the outcome do not have the same consequence on the state as measuring $f(A)$. The quantum analogue of two random variables X and Y will be described as soon as we introduce quantum composite systems.

An important property for the description of noisy quantum information is the fact that the normalization condition can also be written as a trace[5] condition

$$\mathrm{Tr}\,|\psi\rangle\langle\psi| = 1 \tag{2.9}$$

and, assuming normalization, the expectation value can be written as

$$\langle A \rangle = \mathrm{Tr}(A\,|\psi\rangle\langle\psi|). \tag{2.10}$$

---

[4]An analytic function $f(A)$ is a function that can be written as a convergent series $f(A) = \sum f_n A^n$. Thanks to the existence of a spectral decomposition of Hermitian operators $A = \sum_i a_i P_i$, we have $A^n = \sum_i a_i^n P_i$, and this implies $f(A) = \sum f_n A^n = \sum_i f(a_i) P_i$.

[5]The trace of a square matrix is defined as the sum of the elements on the main diagonal and it is independent under a change of orthonormal basis.

It is important to note that the assumption that the vector space is a Hilbert space allows us to choose an orthonormal basis for the vector space. In the finite dimension, a finite orthonormal basis can be chosen and bra-ket operations reduced to operations on complex vectors and matrices, and any element can be written as

$$|\psi\rangle = \sum_{i=1}^{n} c_i |u_i\rangle, \tag{2.11}$$

where n is the dimension of the Hilbert space, $c_i \in \mathbb{C}$ are the complex coefficients, and $\{|u_i\rangle\}$ is an orthonormal basis for the Hilbert space. With this decomposition, the inner product between two vectors $|\chi\rangle$ and $|\psi\rangle$ with coefficients *in the same basis* can be written as

$$\langle \chi | \psi \rangle = \sum_{i,j=1}^{n} d_j^* c_i \langle u_i | u_j \rangle = \sum_{i=1}^{n} d_i^* c_i, \tag{2.12}$$

and the norm becomes

$$\|\psi\|^2 = \sum_{i=1}^{n} |c_i|^2 \geq 0. \tag{2.13}$$

As mentioned previously, complex coefficients c_i are the values that reflect the intrinsic probabilistic nature of quantum mechanics. In fact, they are called probability amplitudes because, *in the presence of an observable* that has eigenvectors $\{|u_i\rangle\}$ *with distinct eigenvalues*

$$A = \sum_{i=1}^{n} a_i |u_i\rangle\langle u_i|, \tag{2.14}$$

the eigenvalues a_i represent the possible outcomes of the measurement, and the squared norms $|c_i|^2 \geq 0$ represent the outcome probabilities of the measured observable, the probabilities of observing the state in a specific basis vector. The normalization condition guarantees that the sum of probabilities is also normalized to one.

It is now possible to define the fundamental block of quantum information, which is called a *qubit*. A qubit is a quantum system described by a two-dimensional Hilbert space $\mathcal{H}^2$, with states $|\psi\rangle$ usually decomposed as

$$|\psi\rangle = c_1|0\rangle + c_2|1\rangle, \tag{2.15}$$

where

$$|0\rangle = \begin{pmatrix} 1 \\ 0 \end{pmatrix} \tag{2.16}$$

$$|1\rangle = \begin{pmatrix} 0 \\ 1 \end{pmatrix} \tag{2.17}$$

represent the elements of the commonly used computational basis. Qubits can be used to describe real physical characteristics such as spin and polarization.

Moreover, a qubit can also be generalized to so-called *qudits*, which are arbitrary d-dimensional quantum systems. Formally, a qudit is a quantum system described by a d-dimensional Hilbert space $\mathcal{H}^d$, with states $|\psi\rangle$ generally decomposed in the computational basis as

$$|\psi\rangle = \sum_{i=1}^{d} c_i |i\rangle, \tag{2.18}$$

where c_i are the amplitudes which satisfy $\sum_{i=1}^{d} |c_i|^2 = 1$, and $\{|i\rangle\}$ is a set of orthonormal basis in $\mathcal{H}^d$.

Finally, in order to define an operator K, it is enough to define how it acts on a basis and extend the definition by linearity. In particular, to define an operator it is always sufficient to state how the operator acts on the computational basis. Therefore it is sometimes customary to write

$$|i\rangle \rightarrow |\psi_i\rangle, \tag{2.19}$$

where the operator can be recovered as

$$K = \sum_i |\psi_i\rangle \langle i|. \tag{2.20}$$

This only works if $|i\rangle$ is an orthonormal basis; if the vectors $|i\rangle$ are not orthogonal to each other, then the operator K defined by Eq. (2.20) is not the same operator defined by Eq. (2.19).

One of the main reasons for the introduction of the bra-ket notation was the recurrent need to identify an orthonormal basis of vectors via the eigenvalues of commuting observables. If an observable X has non-degenerate spectrum/eigenvalues, then the choice of the orthonormal basis that diagonalizes X is unique up to the phases, and the spectrum can be used to index the eigenvectors. Namely, in standard notation, for any eigenvalue in the spectrum $x \in \sigma(X)$, we could denote with $\mathbf{v}_x$ the corresponding eigenvector so that $X\mathbf{v}_x = x\mathbf{v}_x$ is satisfied. In bra-ket notation this becomes $X|x\rangle = x|x\rangle$ for all $x \in \sigma(X)$, which might take some time getting used to.

One way to get accustomed to this use of Dirac's notation may be to initially write $X|X = x\rangle = x|X = x\rangle$ (similar to how it is done for probabilities of random variables), as $|x\rangle$ are indeed the states for which the observable X has a well-defined value, namely x. If the eigenvalues of X are degenerate, then the eigenvalue does not uniquely identify the eigenstate and this method cannot be used. However, if a second observable Y commutes with X, then there exists an orthonormal basis of eigenvectors for both X and Y. More precisely, if $XY = YX$, then there always exists a simultaneous solution $|v\rangle$ to the eigenvalue equations $X |v\rangle = \lambda_X |v\rangle$ and $Y |v\rangle = \lambda_Y |v\rangle$; the eigenvector $|v\rangle$ is then called a joint eigenvector of X and Y (the corresponding eigenvalues do not need to be the same). The observables X and Y might be degenerate but their joint eigenvectors might be unique, in which case the pair of eigenvalues of X and Y can be used to index the joint eigenvectors as $|x, y\rangle \equiv |X = x, Y = y\rangle$, by implicitly defining them as the solution of the *pair* of eigenvalue equations $X |x, y\rangle = x |x, y\rangle$ and $Y |x, y\rangle = y |x, y\rangle$ with $x \in \sigma(X)$ and $y \in \sigma(Y)$.

Mathematically, this style of implicitly defining eigenvectors as solutions to eigenvalue equations might not make a lot of sense. However, physically this has meaning because the eigenvalues are actual possible outcomes of physical experiments.

The typical examples are the states of the electron in the hydrogen atom. The same states, albeit with different energies and occupation order, are used to index the chemical elements in the periodic table. The basis states and the atom elements are indexed by four numbers: $n \in \mathbb{N}$ as the energy units of the state and the row in the periodic table, $l \in \mathbb{N}$ as the magnitude of angular momentum at a given energy and the column groups of the periodic table (physically achieved *stable* values are $0, 1, 2, 3$, commonly referred to with the orbital type letters s, p, d, f), and finally, $m \in \{-l, \ldots, +l\}$, one of the components of angular momentum and $s = \pm\frac{1}{2}$, the spin, which together index the string of chemical elements narrowed down by n and l.

The numbers n, l, m, and s do not index all the possible atomic states, they index an orthonormal basis $|n, l, m, s\rangle$ for the atomic states, implicitly defined as the eigenvectors of the energy, magnitude of angular momentum, component of angular momentum and spin. These observables are by themselves degenerate, but together their eigenvectors are not.

Theorem 2.1 (Heisenberg's Uncertainty Principle) *If we are given two observables, represented by commuting operators A, B, they can be concurrently measured with arbitrary accuracy. On the other hand, if A, B do not commute, then a measurement with arbitrary accuracy is not possible. Especially, given the standard deviations ΔA and ΔB,[6] their product will always exceed a minimum value, which depends on the commutator of the physical observables. Formally, the uncertainty principle can be written as*

$$\Delta A \Delta B \geq \frac{|\langle [A, B] \rangle|}{2}. \tag{2.21}$$

---

[6]The variance ΔX^2 of an observable X is itself defined usually as the expectation value $(X - \langle X \rangle)^2$, simply using the quantum definition of the expectation value. $(X - \langle X \rangle)^2$ is itself a quantum observable with $\langle X \rangle$ being simply the constant observable $\langle X \rangle \, \mathbb{1}$.

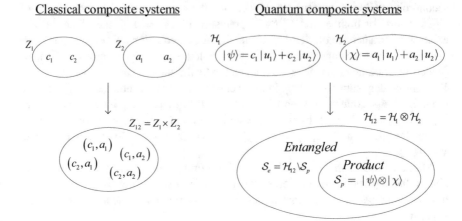

Fig. 2.5 Comparison between the mathematical structure of classical and quantum composite systems

2.1.3   Composite Systems and Entanglement

In reality, it is common to have composite systems instead of single isolated ones. In quantum mechanics, composite systems show some very strange and interesting behaviors. Entanglement is one of these and it was originally identified by Erwin Schrödinger. In fact, he defined it with the German word *Verschränkung* in an article in 1935 (*"Die gegenwiirtige Situation in der Quantenmechanik"*), where he answered Albert Einstein regarding the Einstein–Podolsky–Rosen paradox. Entanglement is an intrinsic aspect of quantum systems. This is a kind of correlation, which has no equivalent in classical mechanics. In particular, two or more arbitrarily distant quantum systems can still form and act as a single entity. An exhaustive survey on entanglement theory is [HHHH09]. Figure 2.5 depicts the mathematical difference between classical and quantum composite systems, by highlighting the presence of entanglement in the latter.

The space of all possible states of a classical composite system can be described by the Cartesian product of the parts which compose the whole system itself (as it happens with Z_1 and Z_2 in the example of Fig. 2.5).

On the other hand, by considering quantum composite systems, the state space of the whole system becomes the tensor product of the state spaces of the components. Formally, given n states $\{|\psi_1\rangle, \ldots, |\psi_n\rangle\}$ in Hilbert spaces $\mathcal{H}_1, \ldots, \mathcal{H}_n$ respectively, the state of the total system becomes the *product state* $|\psi_1\rangle \otimes |\psi_2\rangle \otimes \ldots \otimes |\psi_n\rangle$ in the Hilbert space $\mathcal{H}_1 \otimes \ldots \otimes \mathcal{H}_n$. Figure 2.5 shows the example of two closed quantum systems $|\psi\rangle$ and $|\chi\rangle$ (with their respective coefficients c_1, c_2 and a_1, a_2), composing a system in a product state $|\Psi\rangle \in \mathcal{H}_{12}$. This represents the state of n isolated systems that are brought together. However, the set of all possible composite quantum states $\mathcal{H}_{12}$ is the tensor or Kronecker product space $\mathcal{H}_1 \otimes \mathcal{H}_2$ and thus also contains superpositions (linear combinations) of product states. In particular, it will contain non-product states, which are called entangled states. It is important to notice that the two subsets of product and entangled states are not vector subspaces.

In Fig. 2.5, the example of the quantum composite system represents the space of states of a system of two spins $1/2$. Especially, any product state $|\psi\rangle \otimes |\chi\rangle$ can be written as

$$c_1 a_1 |u_1\rangle \otimes |u_1\rangle + c_1 a_2 |u_1\rangle \otimes |u_2\rangle + c_2 a_1 |u_2\rangle \otimes |u_1\rangle + c_2 a_2 |u_2\rangle \otimes |u_2\rangle \tag{2.22}$$

while an example of an entangled state is the well-known maximally entangled state

$$\frac{1}{\sqrt{2}} \left(|u_1\rangle \otimes |u_1\rangle + |u_2\rangle \otimes |u_2\rangle \right). \tag{2.23}$$

It is extremely important to notice that these vector states are the analog of classical deterministic probability distributions. However, all deterministic probability distributions of composite systems are product distributions. In quantum systems, already in the *deterministic* case we see non-product effects, and thus correlations. As soon as we introduce noisy states, we will also need to update the definition of entanglement in order to distinguish genuine entangled correlations from the classical correlation of non-deterministic probability distributions.

The above is actually a general way of constructing entangled states out of a pair of orthogonal states. Let two unit vectors $|\varphi_1\rangle, |\varphi_2\rangle \in \mathbb{C}^d$ with $|\varphi_1\rangle \perp |\varphi_2\rangle$, and nonzero coefficients $\alpha, \beta \in \mathbb{C}, |\alpha|^2 + |\beta|^2 = 1$ be given. We show, that

$$|\eta\rangle = \alpha |\varphi_1\rangle \otimes |\varphi_1\rangle + \beta |\varphi_2\rangle \otimes |\varphi_2\rangle \tag{2.24}$$

is state vector of an entangled state in $\mathbb{C}^d \otimes \mathbb{C}^d$. In fact, the assumption that η is a product state leads to a contradiction. Let $\{|\varphi_1\rangle, |\varphi_2\rangle, \ldots, |\varphi_d\rangle\}$ be an extension of $|\varphi_1\rangle, |\varphi_2\rangle$ to an orthonormal basis in $\mathbb{C}^d$. If $|\eta\rangle$ was a product state, then it would take the form of a product vector

$$|\eta\rangle = \left(\sum_{i=1}^{d} c_i |\varphi_i\rangle \right) \otimes \left(\sum_{i=1}^{d} \tilde{c}_j |\varphi_j\rangle \right) = \sum_{i,j=1}^{d} c_i \tilde{c}_j |\varphi_i\rangle \otimes |\varphi_j\rangle. \tag{2.25}$$

Comparing the coefficients in Eqs. (2.24) and (2.25) shows that $|\eta\rangle$ would satisfy

$$c_1 \cdot \tilde{c}_1 = \alpha, \quad c_2 \cdot \tilde{c}_2 = \beta, \quad c_i \cdot \tilde{c}_j = 0 \text{ for } i \neq j \text{ or } i, j > 2. \tag{2.26}$$

As a consequence of the above equalities, we would have

$$\alpha \cdot \beta = c_1 \cdot \tilde{c}_2 \cdot \tilde{c}_1 \cdot c_2 = 0 \tag{2.27}$$

which is a contradiction to $\alpha, \beta \neq 0$.

Example 2.1 A pure bipartite state $|\eta\rangle = \alpha \cdot |\varphi_1\rangle \otimes |\varphi_1\rangle + \beta \cdot |\varphi_2\rangle \otimes |\varphi_2\rangle$, $|\alpha|^2 + |\beta|^2 = 1$ is entangled if and only if $\alpha, \beta \neq 0$.

Theorem 2.2 (No-Cloning Theorem) *There does not exist a universal copier[7] of arbitrary quantum states, namely the map $|\psi\rangle \rightarrow |\psi\rangle \otimes |\psi\rangle$ is not a physical map because it is not linear. Thus, arbitrary states cannot be cloned in quantum mechanics.*

Notice that in discussing the entangled state, we have only picked the special case where only one index is used in the decomposition, namely all terms are of the form $|\phi_i\rangle \otimes |\phi_i\rangle$ rather than $|\phi_i\rangle \otimes |\phi_j\rangle$. The reason is the following particularly powerful result, stating that we can always decompose an entangled state into a superposition of terms of the form $|\phi_i\rangle \otimes |\phi_i\rangle$.

Theorem 2.3 (Schmidt Decomposition) *Let $|a\rangle \in \mathcal{H}_1 \otimes \mathcal{H}_2$. Then there exist orthonormal systems $\{|v_i\rangle\}_{i=1}^m \subset \mathcal{H}_1$, $\{|w_i\rangle\}_{i=1}^m \subset \mathcal{H}_2$ and numbers $\alpha_1 \geq \alpha_2 \geq \cdots \geq \alpha_m > 0$, such that*

$$|a\rangle = \sum_{i=1}^m \sqrt{\alpha_i} \, |v_i\rangle \otimes |w_i\rangle \tag{2.28}$$

holds.

The Schmidt decomposition Theorem is essentially a reformulation of the singular value decomposition, because all operators can be mapped into vectors of the tensor product of the input and output Hilbert spaces and vice versa.

Remark 2.1

1. m is usually called the *Schmidt number* of $|a\rangle$, and $\sqrt{\alpha_1}, \ldots, \sqrt{\alpha_m}$ the *Schmidt coefficients* .
2. It holds

$$A_1 := \mathrm{Tr}_{\mathcal{H}_2}(|a\rangle \langle a|) = \sum_{i=1}^m \alpha_i \, |v_i\rangle \langle v_i| \tag{2.29}$$

$$A_2 := \mathrm{Tr}_{\mathcal{H}_1}(|a\rangle \langle a|) = \sum_{i=1}^m \alpha_i \, |w_i\rangle \langle w_i| . \tag{2.30}$$

In particular, the Schmidt coefficients are the nonzero eigenvalues (not necessarily distinct), and the vectors in the orthonormal systems which appear in the Schmidt decomposition are the corresponding eigenvectors.

---

[7] A universal copier is a device that can copy any arbitrary input quantum state.

Though formulated for general bipartite vectors, we will use the Schmidt decomposition most of the time for pure quantum states. Unfortunately, there is no general extension of Theorem 2.3 to spaces with more than two tensor factors. The problem in larger tensor products is known as computing the tensor rank of the vectors and has important applications in both classical computation complexity and quantum entanglement theory.

2.1.4 Composite Observables

We have seen that the pure vector states of composite systems live in the tensor product Hilbert space of the systems. Analogously, the tensor product structure can be extended to linear operators, and observables of composite systems can be found in the tensor product space of linear operators, which is equivalent to the space of linear operators on the tensor product vector space. Namely, let $\mathcal{H}_1, \ldots, \mathcal{H}_N$ be input Hilbert spaces, let $\tilde{\mathcal{H}}_1, \ldots, \tilde{\mathcal{H}}_N$ be output Hilbert spaces, and let $\mathcal{L}(\mathcal{H}_i, \tilde{\mathcal{H}}_i)$ denote the spaces of linear operators. The tensor product

$$\mathcal{L}(\mathcal{H}_1, \tilde{\mathcal{H}}_1) \otimes \cdots \otimes \mathcal{L}(\mathcal{H}_N, \tilde{\mathcal{H}}_N) \simeq \mathcal{L}(\mathcal{H}_1 \otimes \cdots \otimes \mathcal{H}_N, \tilde{\mathcal{H}}_1 \otimes \cdots \otimes \tilde{\mathcal{H}}_N) \tag{2.31}$$

is the space spanned by the operators $A_1 \otimes \cdots \otimes A_N$ defined as

$$(A \otimes \cdots \otimes A_N)(|v_1\rangle \otimes \cdots \otimes |v_N\rangle) = A_1 |v_1\rangle \otimes \cdots \otimes A_N |v_N\rangle. \tag{2.32}$$

The space of observables of $\mathcal{H}_1, \ldots, \mathcal{H}_N$ is the space of Hermitian operators in $\mathcal{L}(\mathcal{H}_1 \otimes \cdots \otimes \mathcal{H}_N) \equiv \mathcal{L}(\mathcal{H}_1 \otimes \cdots \otimes \mathcal{H}_N, \mathcal{H}_1 \otimes \cdots \otimes \mathcal{H}_N)$.

For vectors, this space will contain observables that cannot be written as product observables. However, whether an Hermitian operator is a product does not determine whether the observable is an entangled observable. The situation in this case is more diverse. Let A and A' be observables on $\mathcal{H}$ and $\mathcal{H}'$, respectively. The corresponding observables in the composite system $\mathcal{H} \otimes \mathcal{H}'$ are

$$A \otimes \mathbb{1} \qquad\qquad \mathbb{1} \otimes A' \tag{2.33}$$

which are now the analogue of two random variables X and Y. Notice that these observables always commute, as we expect from observables that do not interact and in particular from observables of distinct systems. However, they still might be correlated because the state might not be a product state (whether entangled or not).

Again, it is important to notice that computing a function of the outcomes $f(a, a')$ after the measurement is not equivalent to measuring $f(A, A')$, as might already be the case for the sum and product observables

$$A \otimes \mathbb{1} + \mathbb{1} \otimes A' \qquad A \otimes A'. \tag{2.34}$$

Moreover, this may not be implementable without interaction and would need to be considered an entangled measurement. Indeed, if $A = \sum_i a_i P_i$ and $A' = \sum_j a'_j P'_j$ are the spectral decompositions of A and A', then the right hand side of

$$A \otimes A' = \left(\sum_i a_i P_i \right) \otimes \left(\sum_i a'_j P'_j \right) = \sum_{i,j} a_i a'_j \cdot P_i \otimes P_j \tag{2.35}$$

is not necessarily the spectral decomposition of $A \otimes A'$. Indeed, the correct spectral decomposition is $A = \sum_k \bar{a}_k \bar{P}_k$ with

$$P_k = \sum_{i,j:\ a_i a'_j = a_k} P_i \otimes P'_j. \tag{2.36}$$

If any of these projectors is not a product, then it can be considered an entangled projector, in the sense that the subspace will contain some entangled state that is left unaffected by the projector and thus the measurement. Such measurements cannot be performed by acting independently on the two subsystems or even coordinating the measurements with classical randomness and communication.

An example is the parity observable $Z \otimes Z$ on two qubits, which has spectrum $\{\pm 1\}$ just like Z. Measuring Z separately implies obtaining two values out of the spectrum, one for each qubit, resulting in projectors $|00\rangle\langle00|$, $|01\rangle\langle01|$, $|10\rangle\langle10|$, and $|11\rangle\langle11|$. The observable parity, however, is equal to

$$Z \otimes Z = (|00\rangle\langle00| + |11\rangle\langle11|) - (|01\rangle\langle01| + |10\rangle\langle10|) \tag{2.37}$$

and thus collapses the state only with projectors $|00\rangle\langle00| + |11\rangle\langle11|$ and $|01\rangle\langle01| + |10\rangle\langle10|$. In particular, the entangled state $c_0 |00\rangle + c_1 |11\rangle$ are eigenvectors of the parity observable and are not disturbed by its measurement, but they are disturbed by the independent Z measurements. $Z \otimes Z$ cannot be measured without a nontrivial interaction with the two qubits.

For an observable to induce a *non-entangled* measurement, it is at least necessary that all the projectors be product projectors. Even then, the measurement might not be implementable as classically coordinated separate measurements [BDF+99]. We need to wait until Chap. 4 for a definition of what measurements can be performed without entanglement.

2.2 Noise in Quantum Systems

Up to now, the discussion has been focused on modeling states of quantum systems via state vectors and braket notation.

However, while classical mechanics is supposed to be a special case of quantum mechanics, vector states and unitary operations clearly cannot model classical operations because they would be too restrictive. Additionally, it should be possible to describe classical mechanics without the need for an ad hoc measurement process. Namely, it should be possible to describe quantum systems and measurements in a unified way because even if a measurement involves hidden interacting systems, by including the hidden systems in the description, the whole process will describe a quantum system without measurement. Said otherwise, we can always avoid a measurement by considering a larger system. This solution is however cumbersome, and in order to overcome this, it is more useful to analyze quantum systems via the so-called *density matrix* formalism. Density matrices are also called either *density operators* or *state operators*. As a bonus, writing the pure states as density matrix also removes the remaining freedom of choice in normalized vector states, where different unit vectors that differ by an arbitrary phase factor e^{iy} still describe the same quantum state.

The new description is equivalent to the description given so far and has the advantage of modeling both noise and classical systems. Thus, without loss of generality, the objects that will be introduced in this section can be taken as equivalent postulates to the formulation or quantum mechanics that we have already seen, and are the starting point of quantum information. The changes from postulates of quantum mechanics to the postulates of quantum information that we will see in this section are briefly summarized in Fig. 2.6.

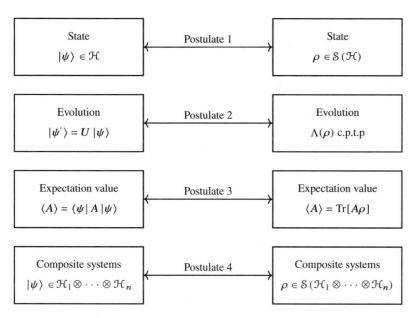

Fig. 2.6 Summary of postulates of quantum mechanics from the point of view of state vectors and density operators

2.2.1 Density Matrix

In closed systems, because the expectation value of an observable can be computed equivalently as $\langle A \rangle = \langle \psi | A | \psi \rangle = \text{Tr}[A | \psi \rangle \langle \psi |]$, a pure state vector $| \psi \rangle$ can equivalently be represented by the corresponding (orthogonal) projector[8] $\mathcal{P}_\psi = | \psi \rangle \langle \psi |$.

If the system is not isolated but is open, namely part of a larger isolated composite quantum system, then the state of the single system cannot be written as a pure state unless the global pure state is a product. That is, the state cannot be written in terms of wave functions or kets. Information on these systems is incomplete and it is not possible to assign proper state vectors to entangled subsystems. In this case, states are still normalized but they do not have to be projectors. Taking advantage of the fact that a basis of the global system $\mathcal{H} \otimes \mathcal{H}'$, constructed as a tensor product of the basis of the subsystem always exists, the expectation values of subsystem observables $A \otimes \mathbb{1}$ can be written as an expectation value over a partially traced state,

$$\text{Tr}\left[(A \otimes \mathbb{1}) | \psi \rangle \langle \psi |\right] = \sum_{i,j} (\langle u_i | \otimes \langle v_j |)(A \otimes \mathbb{1}) | \psi \rangle \langle \psi | (| u_i \rangle \otimes | v_j \rangle)) \tag{2.38}$$

$$= \sum_i \langle u_i | A \left[\sum_j (\mathbb{1} \otimes \langle v_j |) | \psi \rangle \langle \psi | (\mathbb{1} \otimes | v_j \rangle)) \right] \tag{2.39}$$

$$= \text{Tr}\left[A \left(\sum_j | \psi_i' \rangle \langle \psi_i' | \right)\right] \tag{2.40}$$

$$= \text{Tr}[A \, \text{Tr}_{\mathcal{H}'} \, \rho]. \tag{2.41}$$

The partially traced state

$$\rho = \text{Tr}_{\mathcal{H}'} | \psi \rangle \langle \psi | = \sum_j | \psi_i' \rangle \langle \psi_i' | = \sum_j p_i | \psi_i \rangle \langle \psi_i | \tag{2.42}$$

is called a density operator and does not depend on the choice of basis, even though the decomposition does, and still obeys the normalization condition $\text{Tr} \, \rho = 1$. If the initial state $| \psi \rangle$ was not a product, ρ will not be a pure state, in which case it is called a mixed state. The density operator of mixed states not only depends on the projectors of each state $\mathcal{P}_{\psi_i}$ but also on their respective probabilities p_i (see Fig. 2.7).

---

[8]Given a k-dimensional subspace W of the d-dimensional vector space V. Using the Gram–Schmidt procedure, it is possible to construct an orthonormal basis $|1\rangle, \ldots, |d\rangle$ for V such that $|1\rangle, \ldots, |k\rangle$ is an orthonormal basis for W. Then $\mathcal{P} \equiv \sum_{i=1}^k |i\rangle \langle i|$ is the projector onto the subspace W. Orthogonal projectors are an important class of Hermitian operators in quantum theory.

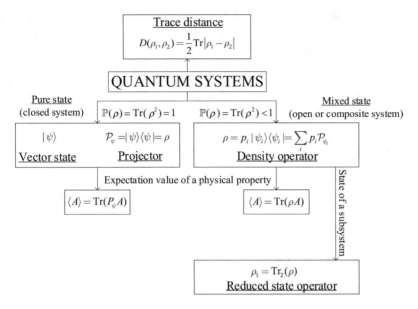

Fig. 2.7 Taxonomy of quantum systems. Closed systems (pure states) can be represented via vectors or density operators (in this case they coincide with projectors) while open systems (also with mixed states) are more effectively described via density operators. By using representation via operators, the expectation value of a property is calculated using trace. Subsystems of composite systems can be described via reduced state operators obtained through partial tracing

A collection of quantum states $|\psi_i\rangle$ with an associated probability distribution p_i is called an ensemble of states. The density operator, or density matrix (in finite dimensions), is defined as the average $\rho = \sum_i p_i \mathcal{P}_{\psi_i}$ of its quantum state projectors over its probability distribution. Every ensemble has a unique density operator but the opposite does not necessarily hold, thus every density operator does not correspond to a unique ensemble and could correspond to many ensembles. Namely, it is also possible for two completely different ensembles to have the same probability of measuring results. The density operator can also be seen as the state of a given quantum system, so it can be used to calculate probabilities for any measurement performed on that system. This also affects the calculation of the expectation value, in which the term of the trace also becomes the weighted sum of the projectors with their respect probabilities.

Every system admits a density operator, and every density operator is a valid state. Thus, in quantum information it is possible, without loss of generality, to replace Postulate 1 and say the set of possible states is the set of all possible density operators defined as

$$\mathcal{S}(\mathcal{H}) = \{\rho \in \mathcal{L}(\mathcal{H}) : \mathrm{Tr}\,\rho = 1,\ \rho \geq 0\}, \tag{2.43}$$

where $\rho \geq 0$ denotes that ρ is Hermitian and has a positive spectrum.

Given the representation of states of quantum systems via density matrices (operators), it is possible to define a quantity that makes it possible to measure if a state is pure or mixed: this measure is called purity (see Fig. 2.7). If the purity $\mathbb{P}(\rho) = \operatorname{Tr} \rho^2$ is one, then the quantum state is pure, while quantum states are mixed for values less than one. If the purity of a density operator is equal to one, the equation $\rho|\psi\rangle = |\psi\rangle$ has a single eigenvector $|\psi\rangle$ and then ρ becomes the projector $\rho = |\psi\rangle\langle\psi|$.

Definition 2.1 (Purification) Let $\rho \in \mathcal{S}(\mathcal{H})$ be a state, and $\mathcal{K}$ be an additional Hilbert space. A pure state $|\Psi\rangle \langle\Psi|$ is a *purification* of ρ if $\operatorname{Tr}_{\mathcal{K}} |\Psi\rangle \langle\Psi| = \rho$.

A purification can always be found from a diagonalization of the state, the square root of the eigenvalues and the eigenvectors always provide a Schmidt decomposition of a vector state in $\mathcal{H} \otimes \mathcal{H}$ that is a purification of ρ. This theorem states that all density matrices can be described as the physical state of a non-isolated system. Therefore we can define all the states, statistically, to be any density matrix without worry, because this definition does not include non-physical density matrices that are not the reduced state of a physical pure state.

2.2.2 The Bloch Sphere of a Qubit

We now focus on a useful representation of the states of a qubit, the quantum counterparts of classical bit systems with the alphabet $|\mathcal{X}| = 2$. The set $\mathcal{S}(\mathbb{C}^2)$ of qubit states has a convenient pictorial representation in $\mathbb{R}^3$, called the *Bloch sphere*, which we derive next. The Bloch sphere is mathematically equivalent to other representations of the unit complex sphere in real space, and in particular it is equivalent to the Poincaré sphere representing the states of light polarization. The matrices

$$\sigma_0 = \begin{pmatrix} 1 & 0 \\ 0 & 1 \end{pmatrix} \qquad \sigma_1 = \begin{pmatrix} 0 & 1 \\ 1 & 0 \end{pmatrix} \tag{2.44}$$

$$\sigma_2 = \begin{pmatrix} 0 & -i \\ i & 0 \end{pmatrix} \qquad \sigma_3 = \begin{pmatrix} 1 & 0 \\ 0 & -1 \end{pmatrix} \tag{2.45}$$

are called *Pauli matrices* and form an orthogonal basis in $\mathcal{L}(\mathbb{C}^2)$. It holds

$$\langle \sigma_i, \sigma_j \rangle_{HS} = \operatorname{Tr} \left(\sigma_i^\dagger \sigma_j \right) 2\delta_{ij}, \tag{2.46}$$

where

$$\langle A, B \rangle_{HS} = \operatorname{Tr} \left(A^\dagger B \right) \tag{2.47}$$

is the Hilbert–Schmidt complex inner product in matrix spaces. If we normalize each of the matrices with a factor $1/\sqrt{2}$ we obtain an orthonormal basis. We can write each matrix $A \in \mathcal{L}(\mathcal{H})$ as a linear combination

$$A = \frac{1}{2}\sum_{i=0}^{3}\langle A, \sigma_i\rangle_{HS}\,\sigma_i = \frac{1}{2}\sum_{i=0}^{3}r_i\sigma_i \qquad (2.48)$$

with $r_i = \langle A, \sigma_i\rangle_{HS}$. We aim to derive conditions on the numbers $r_0, \ldots, r_3$ being equivalent to $\rho \in \mathcal{S}(\mathcal{H})$. We have

1. Each r_i has to be real, because ρ is Hermitian.
2. $r_0 = 1$ holds because of the property $\mathrm{Tr}(\rho) = 1$.
3. Since $\rho \geq 0$ holds,

$$\frac{1}{4}(r_0^2 - r_1^2 - r_2^2 - r_3^2) = \det \rho \geq 0. \qquad (2.49)$$

Since $r_0 = 1$, we obtain the condition $\|r\| \leq 1$ for the vector $r = (r_1, r_2, r_3)^T \in \mathbb{R}^3$.

On the other hand, if ρ is represented as in Eq. (2.48) with $r_0 = 1$ and $r = (r_1, r_2, r_3)^T$ satisfying $\|r\| \leq 1$, then $\rho \in \mathcal{S}(\mathcal{H})$ is implied. Indeed, $1 = r_0 = \mathrm{Tr}\,\rho$, and $\det \rho = 1 - r_1^2 - r_2^2 - r_3^2 \geq 0$. Consequently, ρ is a density matrix. Since the basis coefficients r_0, r_1, r_2, r_3 of an element of $\mathcal{L}(\mathcal{H})$ are unique, the map which connects each density matrix with its Bloch vector $(r_1, r_2, r_3)^T$ is a one-to-one. By the linearity of the Hilbert–Schmidt scalar product, it is clear that this map is also affine. Therefore, we have introduced an affine bijection of $\mathcal{S}(\mathcal{H})$ onto the radius of one Euclidean ball in $\mathbb{R}^3$. By this fact, it is clear that the set of extreme points of $\mathcal{S}(\mathcal{H})$, the pure states, correspond to the set of extreme points of the unit sphere around 0 in $\mathbb{R}^3$.

Remark 2.2 In quantum optics, it is common to specify the polarization preparation of a laser beam by giving the corresponding Bloch vector $r = (r_1, r_2, r_3)^T$.

$$\begin{pmatrix} 1 \\ 0 \\ 0 \end{pmatrix} \quad \text{(linear horizontal)} \qquad \begin{pmatrix} -1 \\ 0 \\ 0 \end{pmatrix} \quad \text{(linear vertical)}$$

$$\begin{pmatrix} 0 \\ 1 \\ 0 \end{pmatrix} \quad \text{(linear } 45°\text{)} \qquad \begin{pmatrix} 0 \\ -1 \\ 0 \end{pmatrix} \quad \text{(linear } -45°\text{)}$$

$$\begin{pmatrix} 0 \\ 0 \\ 1 \end{pmatrix} \quad \text{(right circular)} \qquad \begin{pmatrix} 0 \\ 0 \\ -1 \end{pmatrix} \quad \text{(left circular)}$$

$$\begin{pmatrix} 0 \\ 0 \\ 0 \end{pmatrix} \quad \text{(unpolarized)}.$$

By calculating their Bloch vectors, one can verify that the pure states $P_i := |i\rangle \langle i|$, $i \in \{0, 1\}$ are located at the north and south poles of the Bloch ball. The set of states which lie on the straight line connecting the pure states P_0 and P_1 are parameterized by probability distributions on $\{0, 1\}$,

$$\{P(p) := p(0)P_0 + p(1)P_1\}. \tag{2.50}$$

In fact, each straight line connecting two antipodes on the Bloch sphere can be regarded as a version of the classical bit states.

2.2.3 Composite Systems

Recall that by the postulates of quantum mechanics, the state of composite systems, the unit vectors of the tensor product Hilbert spaces, and the observables are the Hermitian operators in this tensor product space. This leads to both states and observables that in general cannot be written as independent (tensor product) states or observables of the subsystems.

Similarly, the set of states of a composite system with $\mathcal{H}_1, \ldots, \mathcal{H}_N$ subsystems is given by the set of density matrices in the tensor product Hilbert space

$$\mathcal{S}\left(\mathcal{H}_1 \otimes \cdots \otimes \mathcal{H}_N\right). \tag{2.51}$$

The partial trace already introduced in order to compute the reduce states then becomes an important tool to map states from composite systems to *marginal states* in a subsystem. In full generality, for two Hilbert spaces $\mathcal{H}_1, \mathcal{H}_2$, given a trace function $\mathrm{Tr} : \mathcal{H}_2 \to \mathbb{C}$, the partial trace over $\mathcal{H}_2$ is the linear map

$$\mathrm{Tr}_{\mathcal{H}_2} : \mathcal{L}(\mathcal{H}_1 \otimes \mathcal{H}_2) \to \mathcal{L}(\mathcal{H}_1) \tag{2.52}$$

defined on product operators as

$$\mathrm{Tr}_{\mathcal{H}_2}(A_1 \otimes A_2) := A_1 \otimes \mathrm{Tr}(A_2) \tag{2.53}$$

for all $A_i \in \mathcal{L}(\mathcal{H}_i)$ and extending by linearity to $\mathcal{L}(\mathcal{H}_1 \otimes \mathcal{H}_2)$. Whenever the partial trace over a system $\mathcal{H}$ is used, the system $\mathcal{H}$ is said to be *traced out*. The corresponding partial trace over $\mathcal{H}_1$, $\mathrm{Tr}_{\mathcal{H}_1}$ is defined analogously. Note that the definition of partial traces easily extends to composite systems with more than two subsystems. Let N parties (each of them with Hilbert space $\mathcal{K}_i$ assigned), share a system with Hilbert space $\mathcal{K}_1 \otimes \cdots \otimes \mathcal{K}_N$. To apply the partial trace on the j-th system, one uses the definition above with $\mathcal{H}_1 = \bigotimes_{i \neq j} \mathcal{K}_i$, $\mathcal{H}_2 := \mathcal{K}_i$ and so on.

The following important properties are satisfied by the partial trace. Let $A \in \mathcal{L}(\mathcal{H}_1 \otimes \mathcal{H}_2)$ and $B \in \mathcal{L}(\mathcal{H}_1)$, then

1. $\mathrm{Tr}(A) = \mathrm{Tr}(\mathrm{Tr}_{\mathcal{H}_1}(A)) = \mathrm{Tr}(\mathrm{Tr}_{\mathcal{H}_2}(A))$,
2. $\mathrm{Tr}(\mathrm{Tr}_{\mathcal{H}_2}(A)B) = \mathrm{Tr}(A(B \otimes \mathbb{1}_{\mathcal{H}_2}))$,
3. $A \geq 0 \Rightarrow \mathrm{Tr}_{\mathcal{H}_2}(A) \geq 0$,
4. $\rho \in \mathcal{S}(\mathcal{H}_1 \otimes \mathcal{H}_2) \Rightarrow \mathrm{Tr}_{\mathcal{H}_2}(\rho) \in \mathcal{S}(\mathcal{H}_2)$.

Since non-deterministic probability distributions are also not necessarily a product, we cannot use being a product as a property to distinguish entanglement anymore. To solve this, we point out a significant difference between the concept of a bipartite state in classical theory and quantum theory. The objects describing noisy classical systems are probability distributions. In composite systems, product probability distributions describe statistically independent systems. Since all deterministic probability distributions in composite systems are product and statistically independent, any probability distribution in composite systems can be written as convex combinations of product distributions. Clearly this is not possible in quantum theory, with the entangled states being the counterexample. Thus, we can take the above as the distinctive feature of classically behaving noise, and we will identify the density matrices that are behaving classically as those states that can be decomposed as convex combinations of product states.

Definition 2.2 A state $\rho \in \mathcal{S}(\mathcal{H}_1 \otimes \mathcal{H}_2)$ is called

1. *uncorrelated* or a *product state* , if it can be written $\rho = \rho_1 \otimes \rho_2$ for some $\rho_1 \in \mathcal{S}(\mathcal{H}_1)$, $\rho_2 \in \mathcal{S}(\mathcal{H}_2)$.
2. a *separable state*, if it admits a finite ($N \in \mathbb{N}$) convex decomposition of the form

$$\rho = \sum_{i=1,\dots,N} \lambda_i \, \rho_i \otimes \sigma_i \tag{2.54}$$

   with a probability distribution $\{\lambda_i\}_{i=1,\dots,N}$ and states $\rho_i \in \mathcal{S}(\mathcal{H}_1)$, and $\sigma_i \in \mathcal{S}(\mathcal{H}_2)$.
3. *entangled*, otherwise.

Remark 2.3 One could ask about infinite or even uncountable convex combinations of product states. These are identified as separable by means of Caratheodory's theorem. It asserts that for a given convex subset $A \subset \mathbb{R}^d$, each element $x \in A$ can be written as a convex combination of at most $d + 1$ extremal elements in A. Consequently, each separable state can be written as a finite convex combination of not more than $2d + 1$ product states. The extreme points of the set of separable states are all pure product states.

Observation A quantum system in a pure state can only be uncorrelated to the *outside world*, i.e., if $\rho \in \mathcal{S}(\mathcal{H} \otimes \mathcal{K})$ is such that $\mathrm{Tr}_{\mathcal{K}} \rho$ is pure, it is necessarily a product state. □

The above observation also holds for classical statistical theory. However, in quantum information theory, this insight becomes a powerful tool when combined with the possibility of *purifying* quantum systems. The bipartite pure state resulting from purification captures all correlations of the system *to the outside world*. This is a *quantum feature*, since purifying systems is not possible in classical theory.

2.2.4   Quantum Channels

Systems undergo state changes which alter their statistical properties before being observed. Such *re-preparations* can result from an environmental influence (*noise*), or be an effect of an intentional modification (*processing*). In this section, we identify the general class of maps which represent such changes of a system state. First we remember the *evolution* maps encountered in classical information theory.

Example 2.2 In classical information theory, the classical channels making changes on the classical states are usually described by *stochastic matrices*. A stochastic matrix $W : \mathcal{X} \to \mathcal{P}(\mathcal{Y})$ is a matrix $W = (W(y|x))_{x \in \mathcal{X}, y \in \mathcal{Y}}$, such that $W(\cdot|x)$ is a probability distribution on $\mathcal{Y}$ for each $x \in \mathcal{X}$. The set of stochastic matrices is convex with the extremal elements being the permutation matrices. Notice that the identity matrix $\mathbb{1}$ is a stochastic matrix and that $\mathbb{1} \otimes W$ is also stochastic if W is stochastic.

Heuristically speaking, we should demand from a *quantum channel*, that it be linear and it should *map density matrices to density matrices*. But things are not that simple, as we will see. In the next definition, we will use the map

$$\mathrm{id}_{\mathcal{K}} : \mathcal{L}(\mathcal{K}) \to \mathcal{L}(\mathcal{K}), \quad \mathrm{id}_{\mathcal{K}}(A) = A \qquad (A \in \mathcal{L}(\mathcal{K})). \qquad (2.55)$$

Definition 2.3 A linear map $\mathcal{T} : \mathcal{L}(\mathcal{H}) \to \mathcal{L}(\mathcal{K})$ is called

1. *positive* if $\mathcal{T}(A) \geq 0$ for each $A \geq 0$ in $\mathcal{L}(\mathcal{H})$;
2. *completely positive (c.p.)* if the map $id_{\mathcal{L}(\mathbb{C}^l)} \otimes \mathcal{T}$ is positive for each $l \in \mathbb{N}$;
3. *trace preserving* if $\mathrm{Tr}\, \mathcal{T}(A) = \mathrm{Tr}\, A$ for each $A \in \mathcal{L}(\mathcal{H})$.

At first sight, the distinction between positive and completely positive maps made above seems redundant. As demonstrated by the following example, there are indeed maps which are positive, but not completely. Namely $\mathcal{T}$ being positive does not imply that $\mathrm{id} \otimes \mathcal{T}$ is also positive.

Example 2.3 (Partial Transposition) The map $\mathcal{E} : \mathcal{L}(\mathbb{C}^2) \to \mathcal{L}(\mathbb{C}^2)$, $\mathcal{E}(A) := A^T$ (the transposition being according to the canonical orthonormal basis) is positive, but not completely positive. The counterexample input for $\mathrm{id} \otimes \mathcal{T}$ is any entangled pure state.

We note that from an operational point of view, we have to demand complete positivity rather than positivity when defining what a quantum channel is. Consider, for instance, a situation where a system A is processed with a map $\mathcal{T}$ in a lab, while the overall state ρ_{AE} of the composite system AE including an additional environment system is not a product state (i.e., A is an *open system*). If $\mathcal{T}$ is allowed to be a positive, but not a c.p. map, the global resulting state $\rho'_{AE} := \mathrm{id}_E \otimes \mathcal{T}(\rho)$ cannot be a density matrix!

Definition 2.4 (Quantum Channel) A linear map $\mathcal{T} : \mathcal{L}(\mathcal{H}) \to \mathcal{L}(\mathcal{K})$ is called a *quantum channel* if it is completely positive trace preserving (c.p.t.p.). The set of quantum channels with input space $\mathcal{H}$ and output space $\mathcal{K}$ is denoted $\mathcal{C}(\mathcal{H}, \mathcal{K})$.

A first set of prominent examples of quantum channels are

- **Isometric (noiseless) evolutions**. With an isometry $V \in \mathcal{L}(\mathcal{H})$, the map

$$\mathcal{V}(A) := V A V^* \qquad\qquad \forall\, a \in \mathcal{L}(\mathcal{H}) \qquad\qquad (2.56)$$

  is a channel;
- **Partial traces**. The map $\mathrm{Tr}_{\mathcal{K}} : \mathcal{L}(\mathcal{H} \otimes \mathcal{L}) \to \mathcal{L}(\mathcal{H})$ as defined in the previous section is a channel;
- **State preparations**. The map $\mathcal{N}_\sigma : \mathcal{L}(\mathcal{H}) \otimes \mathcal{L}(\mathcal{H} \otimes \mathcal{K})$ that maps $A \mapsto A \otimes \sigma$ is a channel for each state $\sigma \in \mathcal{S}(\mathcal{K})$.

Prominent examples for maps which are **not** channels are

- **Transposition**, which is trace preserving but is not completely positive, see Example 2.3;
- **Universal quantum copying device**. The map $T : \mathcal{L}(\mathcal{H}) \to \mathcal{L}(\mathcal{H} \otimes \mathcal{H})$, $\rho \mapsto \rho \otimes \rho$ is not a channel since it is not linear.

Remark 2.4 The unavailability of a universal quantum copying device has a severe impact on the conception of quantum communication systems. On the one hand, it allows for very powerful protocol, protecting information from eavesdropping. On the other hand, the *Quantum Internet* needs completely novel solutions to the problem of long-distance transmission. Read-and-repeat solutions as performed by current *classical repeater stations* are *impossible machines* for quantum transmission!

At first glance, the condition of complete positivity formulated in Definition 2.4 seems not to be very handy. To check if a map $\mathcal{T}$ is completely positive, one would have to check if $\mathrm{id}_{\mathbb{C}^l} \otimes \mathcal{T}$ is positive for all $l \in \mathbb{N}$, which would be a rather hopeless task. Fortunately, there exist characterizations which are more easy to handle. The first one is given in the next theorem. This, in particular, states that the definition of quantum channels as a c.p.t.p. map is not too general, because they can all be thought of as physical channels.

Theorem 2.5 (Kraus Decomposition) *Let* $n := \dim \mathcal{H}$, $m := \dim \mathcal{K}$. *A linear map* $\mathcal{T} : \mathcal{L}(\mathcal{H}) \to \mathcal{L}(\mathcal{K})$ *is completely positive if and only if it admits a representation of the form*

$$\mathcal{T}(B) = \sum_{k=1}^{N} T_k B T_k^* \qquad\qquad \forall\, B \in \mathcal{L}(\mathcal{H}) \qquad\qquad (2.57)$$

with $T_1, \ldots, T_N \in \mathcal{L}(\mathcal{H}, \mathcal{K})$. *A representation as Eq. (2.57) is always possible, then with* $N \le n^2 \cdot m^2$. *Furthermore,* $\mathcal{T}$ *is t.p. if and only if* $\sum_{k=1}^{N} T_k^* T_k = \mathbb{1}_n$ *holds.*

Next, we give a second characterization of completely positive maps. While the Kraus decomposition is a representation *on a lower layer* (a map between matrix spaces is represented by a matrix sum) the theorem below gives a representation on a *higher layer*. A c.p. map is represented by a partial evolution derived from an isometric evolution on a larger space.

Theorem 2.6 (Stinespring Dilation) *Let $\mathcal{T} \in \mathcal{C}(\mathcal{H}, \mathcal{H}')$ be a c.p. map. There exists a representation of $\mathcal{T}$ in terms of a linear map $V : \mathcal{L}(\mathcal{H}) \rightarrow \mathcal{L}(\mathcal{H}' \otimes \mathcal{K})$, with additional Hilbert space $\mathcal{K}$ such that*

$$\mathcal{T}(A) = \text{Tr}_{\mathcal{K}}(V A V^*) \qquad\qquad \forall\, A \in \mathcal{L}(\mathcal{H}) \qquad\qquad (2.58)$$

holds. Furthermore, $\mathcal{T}$ is c.p.t.p. if and only if V is an isometry.

We define quantum channels statistically, in the sense that we define them simply as maps that take density matrices to density matrices, without worrying about whether such constraints always describe possible physical evolutions. The above theorem gives an analog to the purification theorem for density matrices. It states that we can always find an environment, where the evolution can be described physically with a unitary or an isometry as a noiseless evolution, so that the quantum channel is simply the result of losing such an environment.

2.3 Measurements

A measurement is the observation of the value a specific observable (physical quantity) can assume. Contrary to the classical interpretation of nature and communications, in general in quantum mechanics it may not be possible to say that an event has happened or that an observable owns a value before the observer has interacted with the system itself. Thus, measurement is an interaction with the quantum system under consideration that also changes and evolves the quantum-mechanical system itself. However, this evolution in general cannot be correctly described by Schrödinger's equation since this mathematical relation only models closed quantum systems.

Section 2.2.1 introduced the difference between closed and open quantum systems. When a property of a closed quantum system has to be measured, this system interacts and becomes coupled to the other external systems, so that the evolved original system becomes open. Measurement equipment primarily represents this external physical system. Figure 2.8 describes the main models used to study and analyze measurements of observables in quantum-mechanical systems.

Since a closed system may become open under measurement, we might be tempted to think that we can recover a closed system by considering the composite system that includes the measurement instrument. In theory this would be possible if the joint system was isolated, but it is generally impossible in practice. If the joint system with the instrument was isolated, we would expect the instrument giving the outcome to behave as a quantum-mechanical pure state, while what we obtain in practice is a system behaving classically. We can understand this

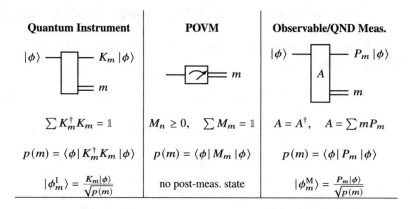

Zurek's Model

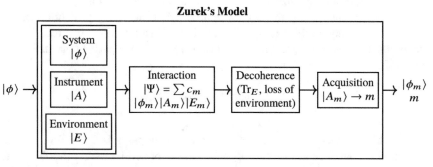

Fig. 2.8 Models to interpret the behavior of measurements on quantum systems

behavior by assuming that the measurement instrument is itself always coupled to an environment which is always lost. This model to describe measurement is sometimes called *Zurek's model* [AFP09] and the environmental process at the basis of this theory is called *decoherence*. Decoherence is the loss of *coherence*, which is the superposition of the states that are to be measured.

The measurement process in Zurek's model is schematized in Fig. 2.8. Formally, a measurement implies not only a coupling between an original quantum system $|\psi\rangle$ and apparatus $|A\rangle$ (as von Neumann originally stated) but also a coupling with the environment $|E\rangle$ (considered a quantum-mechanical system). From this perspective, it can also turn out that any quantum system is, in reality, an open system, because it does not exist in a completely isolated quantum system. After coupling the joint state $|\psi\rangle|A\rangle|E\rangle$ with an interaction, the overall composite open system to be studied becomes

$$|\Psi\rangle = \sum_m K_m |\phi\rangle |A_m\rangle |E_m\rangle = \sum_m \sqrt{p_m} |\phi_m\rangle |A_m\rangle |E_m\rangle , \qquad (2.59)$$

where $\{|A_m\rangle\}$ and $\{|E_m\rangle\}$ are orthonormal bases of the measurement instrument and the environment. If at this point we traced both the measurement instrument and the environment, K_m would indeed be a set of Kraus operators of the noisy channels experienced by the system.

Next, it becomes convenient to represent $|\Psi\rangle$ via its density matrix $|\Psi\rangle\langle\Psi|$, as previously described in Sect. 2.2.1:

$$|\Psi\rangle\langle\Psi| = \sum \sqrt{p_m p_n} \, |\phi_m\rangle \, |A_m\rangle \, |E_m\rangle \, \langle\phi_n| \, \langle A_n| \, \langle E_n| . \tag{2.60}$$

The procedure of measurement reduces the coherence and thus the quantum correlation. The coherence is the information of the superposition in $|\Psi\rangle$ and more precisely, the information of the superposition between the m-th and the n-th term with $m \neq n$ being contained in the term $|\phi_m\rangle \, |A_m\rangle \, |E_m\rangle \, \langle\phi_n| \, \langle A_n| \, \langle E_n|$ and its Hermitian conjugate, which are the off-diagonal elements of $|\Psi\rangle\langle\Psi|$ in this measurement basis. This loss of coherence is mathematically translated into these off-diagonal terms tending to zero. Since both $\{|A_m\rangle\}$ and $\{|E_m\rangle\}$ are orthonormal bases contributing to $|\Psi\rangle$ in a Schmidt decomposition form, the role of the environment is indeed representing this decoherence via the partial trace Tr_E. This results in

$$\sum_m p_m \, |\phi_m\rangle \, |A_m\rangle \, \langle\phi_m| \, \langle A_m| , \tag{2.61}$$

where the coherence between each m-th and n-th term is lost. That is, the information contained in the off-diagonal elements of density matrix $|\Psi\rangle\langle\Psi|$ is left in the environment and lost from the observers point of view. This procedure results in the mixed state, which is finally measured to extract the values m from the state $|A_m\rangle$ of the measurement instrument

$$\sum_m p_m \, |\phi_m\rangle\langle\phi_m| \otimes |m\rangle\langle m| . \tag{2.62}$$

Since the initial interaction is not perfect in reality, namely the state of the system might end up in a slightly different state $|\phi_m\rangle$, the off-diagonal elements are not completely suppressed. As an ideal limit, perfect decoherence is often considered, meaning there are zero-value off-diagonal elements in the intended measurement basis.

The decoherence model is an approach to the measurement problem, which does not introduce external elements to the quantum-mechanical domain (versus, e.g., the von Neumann measurement). However, all measurement models are equivalent and are in agreement with experimental results. Furthermore, the environment is usually inaccessible and thus is not something we want to model explicitly, which was the reason for introducing density matrices and quantum channels. We can think of Zurek's model as a justification for the use of the other models, even if they involve unintuitive interpretations like the wavefunction collapse.

Remark 2.5 Measurements in quantum mechanics can be seen as a decision process [AFP09, Chapter 9] since it is a decision to couple the apparatus and system to be measured, while the environment is involved in the decoherence of the density matrix.

A unified description of all the measurement models, without invoking any explicit environment, is provided by the notion of a *quantum instrument* [DL70]. We can regard a quantum instrument as the map of Eq. (2.62) emerging from Eq. (2.59) on the input state after the loss of the environment

$$\Lambda(\rho) = \sum_m K_m \rho K_m \otimes |m\rangle\langle m| , \tag{2.63}$$

namely a channel with two output systems, i.e., the quantum system containing the post-measurement state and a classical system containing the measurement outcome (recall that the standard basis can be used to model the classical states). The most general form of a quantum instrument (independent of any representation of the linear maps) is actually

$$\Lambda(\rho) = \sum_m \Lambda_m(\rho) \otimes |m\rangle\langle m| , \tag{2.64}$$

where Λ_m are completely positive maps summing to a quantum channel, namely such that $\sum_m \Lambda_m$ is a quantum channel. We can regard Λ_m as partial channels. Either by tracing out the quantum system or by transforming the output into an ensemble of states, we can obtain all the most common measurement models as illustrated in Fig. 2.8. Notice that by tracing the classical system we recover the general description of quantum channels via Kraus operators.

By tracing the quantum part we obtain the most general description of a measurement as a map from a quantum to a classical system without regard for the post-measurement state. This leads to the concept of the *positive operator-valued measure* (POVM). For simplicity let us consider the simpler case of an instrument of the form

$$\Lambda(\rho) = \sum_m K_m \rho K_m \otimes |m\rangle\langle m| , \tag{2.65}$$

which is enough to obtain the general definition of POVM. If we trace the quantum part as we mentioned before, we obtain

$$\text{Tr}_A \Lambda(\rho) = \sum_m \text{Tr}\left(K_m^\dagger K_m \rho\right) |m\rangle\langle m| = \sum_m p_m |m\rangle\langle m| , \tag{2.66}$$

where by construction, $p_m = \text{Tr}[K_m^\dagger K_m \rho]$ must form a probability distribution. All the p_m are positive because the $K_m^\dagger K_m$ are positive, and the restriction that the total trace must sum to one imposes $\sum_m K_m^\dagger K_m = \mathbb{1}$. This leads us to the following definition.

Definition 2.5 Let $\mathcal{H}$ be a finite-dimensional Hilbert space, and $\mathcal{M}$ be a finite set. A POVM (positive operator-valued measure) on $\mathcal{H}$ with measurement outcomes in $\mathcal{M}$ is a family $\{M_m\}_{m \in \mathcal{M}}$ satisfying

1. positivity: $0 \leq M_m \leq \mathbb{1}_{\mathcal{H}}$ for all $m \in \mathcal{M}$
2. completeness: $\sum_{m \in \mathcal{M}} M_m = \mathbb{1}_{\mathcal{H}}$.

The special case of a family of mutually orthogonal projections in $\mathcal{H}$ is called *projection valued measure (PVM)* and if the projectors have all rank one, then it is a *von Neumann measurement* .

The elements M_m are called POVM elements or measurement operators, and sometimes *effects* [AFP09] when using the unifying view of statistical theories, as we will see in the next section. Any positive operator M satisfying $0 \leq M \leq \mathbb{1}$ automatically defines a binary POVM as $\{M, \mathbb{1} - M\}$ and thus the term *measurement operator* can be used to describe any operator satisfying this condition. Given a POVM as defined above, the probability of obtaining a measurement outcome m is given by

$$p(m) = p(m|\rho) = \mathrm{Tr}\left(M_m \rho\right). \tag{2.67}$$

The positivity of both the measurement operator M_m and the state ρ guarantees that $p(m) \geq 0$ and the last condition guarantees $\sum p(m) = \mathrm{Tr}\left(\sum M_m \rho\right) = \mathrm{Tr}\left(\mathbb{1}\rho\right) = 1$.

POVMs satisfy a nice composition relation. Performing two measurements with POVMs $\{M_n\}$ and $\{N_n\}$ sequentially, namely producing $|\phi_m\rangle = M_m |\phi\rangle / p(m)$ and then producing $|\phi_{mn}\rangle = N_n |\phi_m\rangle / p(n|m)$, is equivalent to directly performing the measurement with POVM

$$\{N_n M_m\},$$

meaning that the state satisfies $|\phi_{mn}\rangle = N_n M_m |\phi\rangle / p(m, n)$.

The POVM is a mathematical formalism used in cases where the aspect of interest is only the probability of measuring an outcome and not the state after measurement (i.e., experiments where at the end of the experiment the system requires a single measurement only).

A POVM measurement allows a non-ideal evaluation of incompatible observables, while a PVM $\{P_m\}$ is the probabilistic part of a *projective measurement*, describing the measurement of the observable $H = \sum_m m P_m$, a Hermitian operator on the state space of the system being observed with spectrum $\mathcal{M}$. Sometimes PVMs and POVMs are referred to as the measurement of sharp and unsharp observables [AFP09], with sharp observables coinciding with the Hermitian observables we consider here.

The possible outcomes of the measurement correspond to the eigenvalues m of the observable. Then, by measuring a state ρ, the probability of getting result m is $p(m) = \text{Tr}[P_m \rho]$. Given that outcome m occurred, the state of the quantum system immediately after the measurement becomes $\rho_m P_m \rho P_m / p(m)$. The projectors P_m are also the Kraus operator of the quantum instrument for the measurement in Eq. (2.63) and the procedure described above simply constructs an ensemble of states

$$\{p(m), \rho_m \otimes |m\rangle\langle m|\} = \left\{ p(m), \frac{1}{p(m)} P_m \rho P_m \otimes |m\rangle\langle m| \right\} \tag{2.68}$$

for the output of this quantum instrument. Notice that the collapse induced by the projectors only appears through the conditioning necessary to make an ensemble interpretation out of the quantum-instrument description of the measurement.

Quantum-mechanically, PVMs are a natural description of the value and probability of measuring a certain observable. However, the difference between a POVM and a PVM can be substantial in practice. Imagine a detector that absorbs a photon and correctly gives a signal. This can be regarded as a demolition measurement on the photon, because its state is erased. This process can be modeled with a POVM or quantum instrument, but not with a projective measurement. A projective measurement in this case would be measuring the number of photons in an incoming beam. This has to be done without interfering with the path of the photons in the beam, which would have to be allowed to continue after their number has been determined. Experimentally, this is known as a Quantum-Non-Demolition (QND) measurement, it often includes all the encountered observable measurements and can be extremely challenging to implement physically. It is called *non-demolition* because the state after the measurement is indeed still an eigenstate of the observable (or mixture of eigenstates) with the same eigenvalue and not some other state. The example also illustrates that in general, QND/PVM measurements cannot be performed directly, but rather another quantum system must act as a probe, interacting with the systems to be measured first so that the probe can be measured later while the original system continues to evolve.

An important example of a projective instrument using a probe is the *interaction-free* or *negative-result measurement*, which tries to measure specific properties of a quantum system without interacting with it. Such type of measurements can detect an obstacle with only a small interaction. In some cases, by increasing the repetitions and finesse of the measurement, the probability of success can be sent to 1, without incurring any change in the state [AFP09]. This effectively implements a projective measurement in the limit that detects the obstacle without interacting with it (the projectors onto the Hilbert subspace describe the obstacle and the orthogonal subspace describe its absence).

In general, the structure of Zurek's model tells us that we can always think of a POVM as emerging from a PVM acting on a probe system and vice versa.

We give a final note on von Neumann measurements, where all measurement operators are rank-one projectors, namely projectors onto a basis of pure states.

These measurements are associated with measuring non-degenerate observables, where all the eigenvalues are unique. Such a measurement also correctly estimates the expectation value of any observable diagonal in this basis even though said observables might be degenerate. These observables share the property that they all commute with each other, namely their Lie bracket, $[A, B] = AB - BA$, is zero.

Conversely, doing a sequential measurement of all the observables that are diagonal in the same basis defines a von Neumann measurement. Mathematically, two observables (Hermitian operators) M and N that can be diagonalized in the same basis are called simultaneously diagonalizable, and this can happen if and only if they commute. A basis uniquely defines a class of operators that all commute with each other and, conversely, it is always possible to find a set of commuting operators that uniquely identify a basis.

Quantum-mechanically, a set of commuting observables that uniquely identifies a basis is called a complete set of observables. More concretely, let M and N be two commuting observables. Each observable will have its own spectral decomposition $M = \sum m M_m$, $N = \sum n N_n$ where M_m and N_n are projectors. If the observables commute, then the spectral projectors also commute, and the composed POVM $\{N_n M_m\}$ is also a projective POVM, namely the measurement operators $N_n M_m$ will be a larger set of smaller projectors representing a finer measurement. By choosing more observables in a suitable way, the resulting POVM can be made into a rank-one projective POVM. This is the case of the previously explained example of the energy level (basis states) of the hydrogen atom, determined by the observables: energy, magnitude of angular momentum, one component of angular momentum and spin.

2.4   Quantum Information

In a world where nothing is uncertain, there is no information theory. More precisely, information theory finds its place only because noise and lack of knowledge exist and can be modeled precisely. This is also reflected in the fact that entropy, arguably one of the most important measures of information, is zero for deterministic probability distributions. Communicating something that has a deterministic probability distribution carries no new information (there is no surprise) because the message could have been predicted with certainty.

Noise and uncertainty arise physically as the partial description of a system. It should therefore be no surprise that we can speak of quantum information and its theoretical limit in communication, just like in the classical setting. More concretely, in quantum information theory there exist the same statistical quantities encountered in classical information theory. At the same time, classical information theory cannot describe quantum information and thus some of its concepts must be specific to classical theory while others must be general concepts describing information in any theory. For the sake of completeness and in an attempt to partly demystify quantum mechanics, we will see that both classical and quantum theory fall within a broad concept of statistical theory.

2.4.1   Statistical Theories

Here, no prior familiarity with quantum theory is assumed. Rather, we give a definition of statistical theory that fits both the noisy theory of quantum mechanics and what we call classical statistical theory. We already implicitly use such a *statistical theory* or *generalized probabilistic theory (GPT)* when considering classical Shannon theory.

The statistical structure of quantum theory arises from the evidence that microscopic objects (e.g., atoms, photons,...) tend to exhibit "random behavior" in experiments. Measurement outcomes fluctuate. However, the following assumption (here formulated for the case of finitely many possible measurement values) seems to be valid.

If a measurement is performed on independent and equally prepared systems in the same condition many times, the following relative frequencies converge to *probabilities*:

$$f_i := \frac{\text{number of occurrences of the measurement value } i}{\text{number of total measurements}}. \tag{2.69}$$

The above assumption characterizes the hard core of a statistical theory. We will see that there are additional properties a statistical theory should have, and of course we need to specify, what *measurement*, *independent*, and *equally prepared* means.

It turns out, that the usual *probability theory* used for describing random experiments with coins (and more complex situations) does not suffice to correctly model some experiments with systems like atoms or photons. In this section, we introduce a framework which formulates the specifications of *statistical theory*—in its generality, sufficient to describe the classical and quantum theories.

A real-world statistical experiment is usually divided into two steps.

1. **Preparation:** Setting a preparation P to fix the *initial conditions* of the experiment. Examples:

   - producing dice, throwing the dice,
   - filling an urn with a number of balls, each having a letter from $A \ldots Z$, blindly take a ball from the urn,
   - setting up a certain laser configuration for single-photon production imprinted.

2. **Observation:** Setting an observation R, i.e., rules how observations are made. Here we assume the most basic type of observation—a *yes/no*—measurement: the observation where one of two given alternatives is taking place. For example:

   - observing if the numbers showing up on the dice are odd or even,
   - observing whether or not the letter A is printed on the ball,
   - setting up a semitransparent plate in the laser beam and recording whether it went through the plate or not.

Usually, in a statistical context, the experiment is performed many times (let us say a large number N times). The next step is calculating **relative frequencies**. If N_i is the number of observations of the ith alternative, the relative frequencies are

$$f_1 = \frac{N_1}{N} \quad \text{and} \quad f_2 = \frac{N_2}{N} = 1 - f_1. \tag{2.70}$$

Having the statistical postulate from the preceding paragraph in mind, we assume that the relative frequencies converge in the limit $N \to \infty$.

At this point an important assumption is made: there is a number $p(P, R)$—the *probability* of observation. In order to theoretically describe a certain type of experiment, we form equivalence classes, i.e., we say two preparation procedures P_1 and P_2 are equivalent if they lead to the same probabilities for all observation procedures, i.e., $p(P_1, R) = p(P_2, R)$ for all possible R. Two observation procedures R_1, R_2 are equivalent if $p(P, R_1) = p(P, R_2)$ for all preparation procedures P. We call each equivalence class of preparation procedures a *state*, and each equivalence class of observation procedures an *effect*.

1. **Theoretical description:** A statistical theory is given by a set $\mathcal{S}$ of states and a set $\mathcal{E}$ of effects together with a map $p : \mathcal{S} \times \mathcal{E} \to [0, 1]$, $(s, E) \mapsto p(s, E)$ with

$$p(s_1, E) = p(s_2, E) \text{ for all } E \in \mathcal{E} \Rightarrow s_1 = s_2 \tag{2.71}$$

$$p(s, E_1) = p(s, E_2) \text{ for all } s \in \mathcal{S} \Rightarrow E_1 = E_2. \tag{2.72}$$

Then, we denote *the certain* effect by $\mathbb{1}$, namely $\mathbb{1}$ is the unique effect with $p(s, \mathbb{1}) = 1$ for all states.

We can now write quantum and classical theories as examples of statistical theories describing experiments with a *quantum system* or a *classical system*.

Example 2.4 (Classical (Finite) Statistical Theory) Let Ω be a finite sample set. The statistical theory usually imposed is given by the state set $\mathcal{S} := \mathcal{P}(\Omega)$ and effect set $\mathcal{E} := \mathcal{E}(\Omega)$, with

$$\mathcal{P}(\Omega) := \left\{ q : \Omega \to [0, 1] : \sum_{\omega \in \Omega} q(\omega) = 1 \right\} \tag{2.73}$$

$$\mathcal{E}(\Omega) := \{ f : \Omega \to [0, 1] : 0 \le f(\omega) \le 1 \text{ for all } \omega \in \Omega \}. \tag{2.74}$$

The probability of observing the "fuzzy event"[9] $E \in \mathcal{E}(\Omega)$ (effect) when preparing the probability distribution $q \in \mathcal{P}(\Omega)$ (state) is

---

[9]The reader may be more acquainted with a somewhat restricted model. Traditional (Kolmogorov) probability considers the indicator functions $\mathbb{1}_A$ (for all $A \subset \Omega$) which form a proper subset of $\mathcal{E}(\Omega)$. However, note that these do not form a convex set for nontrivial Ω (cf. the subsequent section.)

$$p(q, E) := \sum_{\omega \in \Omega} q(\omega) \cdot f(\omega). \tag{2.75}$$

As we have seen, the objects defining probabilities in quantum theory are measurement operators and these are exactly the definition of effects in quantum statistical theory.

Example 2.5 (Quantum Statistical Theory) Let $\mathcal{H} = \mathbb{C}^d$ be a finite-dimensional Hilbert space ($d < \infty$). The quantum statistical theory for $\mathcal{H}$ is given by

$$\mathcal{S}(\mathcal{H}) := \{\rho \in \mathcal{L}(\mathcal{H}) : \rho \geq 0 \wedge \operatorname{Tr} \rho = 1\} \qquad \text{(density matrices)} \tag{2.76}$$

$$\mathcal{E}(\mathcal{H}) := \{E \in \mathcal{L}(\mathcal{H}) : 0 \leq E \leq \mathbb{1}\} \qquad \text{(measurement operators).} \tag{2.77}$$

For given density matrix ρ (state) and measurement operator E (effect), the probability of observation is calculated via the formula

$$p(\rho, E) = \operatorname{Tr} \rho E. \tag{2.78}$$

Recall that we associated each effect with a binary observation. The complementary effect $\bar{E}$ is the effect such that for every state s

$$p(s, \bar{E}) = 1 - p(s, E). \tag{2.79}$$

For the classical theories, the certain effect is the indicator function $\mathbb{1}_\Omega$ on the whole event set, while for the quantum theory, it is the identity operator $\mathbb{1}_{\mathcal{H}}$. The complementary effects are then the indicator function on the complementary event set $\mathbb{1}_{\bar{A}}$ for the classical theory, $\mathbb{1}_\Omega - f$ for the fuzzy theory, and $\mathbb{1}_{\mathcal{H}} - E$ for the quantum theory. Notice that in the quantum theory, $\{E, \mathbb{1}_{\mathcal{H}} - E\}$ forms a binary POVM.

2.4.1.1   Convexity

When conducting statistical experiments, there is also the possibility to *mix* preparation procedures as well as observation procedures. For example, when having two devices preparing states s_1, s_2 at hand, one could let a random number generator decide which one of these to take for the next sample. It makes sense to demand that this result is included in an allowable preparation procedure that coincides with the average probability of the outcome for all observations. Namely, the corresponding state $\tilde{s}$, if λ is the probability of preparing s_1, must fulfill

$$p(\tilde{s}, E) = \lambda \cdot p(s_1, E) + (1 - \lambda) \cdot p(s_2, E) \tag{2.80}$$

for every effect E. For both the classical and the quantum theories, p is a linear function of both states and effects, implying that equation i.e., the state corresponds to a mixture of states s_1 and s_2, with the mixed parameter λ being the convex combination.

$$\tilde{S} = \lambda s_1 + (1 - \lambda)s_2, \tag{2.81}$$

i.e., the state corresponding to a mixture of states s_1 and s_2 with mixing parameter λ is their convex combination. A similar argument can be drawn for mixtures of effects and thus both the states S and the effects $\mathcal{E}(\mathcal{H})$ are convex subsets of a linear space. Indeed, probability distributions are vectors and classical channels are linear operators on probability distributions.

This also gives a uniform description of deterministic probability distributions and pure quantum states. Every convex set has a subset of the extreme elements, namely the elements which cannot be written as nontrivial convex combinations of other elements of that set. The extreme elements of the states S are called *pure states* (accordingly, states which are not pure are called *mixed*), while the extreme elements of the effects $\mathcal{E}$ are called *propositions*. For the classical theories, the pure states are the deterministic probability distributions, while the propositions are the indicator functions on a single element. In the quantum theories, the pure states coincide with our original definitions, namely the rank-one projectors stemming from pure vector states, while the propositions are the orthogonal projectors (not necessarily of rank one).

To describe statistical experiments with more than two outcomes, we introduce the concept of an observable. An *observable* (or *measurement*) with a (finite) set X of measurement outcomes is a function $E : X \rightarrow \mathcal{E}$, such that

$$\sum_{x \in X} E_x = \mathbb{1}. \tag{2.82}$$

Since the set of measurement values is finite, it is more common to define a measurement by a collection of effects. In quantum theory this leads us to a general POVM.

2.4.2 Distance Measures

In classical information theory there exist statistical measures of how distinguishable two distributions are from one another. One typical example is the statistical distance, which also finds its counterpart in the quantum setting, the trace distance. However, before we introduce the trace distance, we introduce another equivalent distinguishability measure that is more natural in the quantum setting.

In any inner-product space, the inner product between two vectors is a measure of the overlap between them. Indeed, orthogonal vectors will have zero inner product, while for parallel vectors the inner product will be proportional to the product of the magnitudes. If we restrict to unit vectors, the inner product is an accurate measure of the orthogonality between them.

For quantum states the inner product gains additional meaning, because as we have seen, orthogonal states represent states that can model classical information and can be cloned. Additionally, parallel unit vectors only differ by a phase and are thus the same quantum vector state. The fidelity between two pure states is thus defined as the magnitude square of their inner product

$$F(|\Psi\rangle\langle\Psi|, |\phi\rangle\langle\phi|) = |\langle\psi, \varphi\rangle|^2 = \text{Tr}\left(|\Psi\rangle\langle\Psi| \cdot |\phi\rangle\langle\phi|\right). \tag{2.83}$$

This is zero for orthogonal and distinguishable states, and one only if the states are the same, indicating larger distinguishability as the fidelity becomes smaller. The generalization to mixed states is obtained as follows, which classically, meaning commuting A and B, is known as the square of the Bhattacharyya coefficient. Indeed, there are two concurrent definitions of the quantum fidelity in the literature, sometimes defined as the square root of the following definition.

Definition 2.6 (Quantum Fidelity) Let A, B be positive operators ($A, B \in \mathcal{L}(\mathcal{H})$ and $A, B \geq 0$). The (quantum) fidelity of A and B is defined by[10]

$$F(A, B) := \|A^{\frac{1}{2}} B^{\frac{1}{2}}\|_1^2 = \left(\text{Tr}|A^{\frac{1}{2}} B^{\frac{1}{2}}|\right)^2. = \left(\text{Tr}(A^{\frac{1}{2}} B A^{\frac{1}{2}})^{\frac{1}{2}}\right)^2. \tag{2.84}$$

The fidelity satisfies the following properties

1. $F(A, B) = F(B, A)$,
2. $F(\lambda A, B) = \lambda F(A, B)$ for all $\lambda \geq 0$,
3. $F(|\psi\rangle\langle\psi|, B) = \langle\psi, B\psi\rangle$ for all $\psi \in \mathcal{H}$, and
4. $0 \leq F(A, B) \leq 1$ for density matrices A, B.

The following theorem importantly connects the fidelity of mixed states to the fidelity of pure states via the purification. It rephrases the fidelity as resulting from an optimization of the overlaps of purifications of the states.

Theorem 2.7 (Uhlmann's Theorem) Let $\rho, \sigma \in \mathcal{S}(\mathcal{H})$. It holds

$$F(\rho, \sigma) = \max\left\{|\langle\psi, \varphi\rangle|^2 : \psi \text{ purifies } \rho, \varphi \text{ purifies } \sigma\right\}. \tag{2.85}$$

Notice that the maximization is necessary since we can always construct two purifications that are orthogonal by doubling the dimension of the purifying system and using orthogonal subspaces for the two purifications.

Most important for quantum information is determining a distance measure for channels, as indeed the goal of channel coding is to transform a given channel into an approximation of another, ideally a noiseless channel. From the distance measure

---

[10]Given a matrix A, $|A| = \sqrt{A^\dagger A}$.

of states we derive the distance measure for channels, and the measure of quantum channels induced by the fidelity is known as entanglement fidelity.

Definition 2.7 (Entanglement Fidelity of a Channel) Let $\rho \in \mathcal{S}(\mathcal{H})$ be a state and $\mathcal{N}$ a quantum channel from $\mathcal{H}$ to $\mathcal{H}$. The entanglement fidelity of $(\rho, \mathcal{N})$ is defined by

$$F_e(\rho, \mathcal{N}) := F(|\psi\rangle \langle\psi|, \mathrm{id}_{\mathcal{H}} \otimes \mathcal{N}(|\psi\rangle \langle\psi|) = \langle\psi| \, \mathrm{id}_{\mathcal{H}} \otimes \mathcal{N}(|\psi\rangle \langle\psi|) \, |\psi\rangle, \tag{2.86}$$

where $|\Psi\rangle\langle\Psi|$ is any purification of ρ. Alternatively, the entanglement fidelity can be computed as

$$F_e(\rho, \mathcal{N}) = \sum_{k=1}^{N} |\mathrm{Tr}(A_k \rho)|^2 \tag{2.87}$$

for any Kraus decomposition $\mathcal{N}(\rho) = \sum_{k=1}^{N} A_k \rho A_k^\dagger$, thus showing that the definition is well defined, as it depends neither on the choice of purification for the state nor on the choice of Kraus representation for the channel.

As we have already mentioned, another meaningful measurement is the *trace distance*

$$D(\rho, \sigma) = \frac{1}{2} \mathrm{Tr} \, |\rho - \sigma| \tag{2.88}$$

which measures the physical difference between two quantum states ρ, σ. The unitary transformation of density matrices does not change the trace distance between states. This is the quantum equivalent of the classical statistical distance and has the same interpretation. A measurement can be found that distinguishes the two states with error probability equal to the trace distance. Similar extensions to channels as exist for the fidelity also exist for the trace distance. Actually, trace distance and fidelity are effectively interchangeable only at the cost of losing a square root factor in the distance when transitioning from one to the other [FVDG99]:

$$1 - \sqrt{F(\rho, \sigma)} \leq D(\rho, \sigma) \leq \sqrt{1 - F(\rho, \sigma)}. \tag{2.89}$$

One of the most important properties for a meaningful distinguishability measure is a *data processing inequality*. Namely, the distinguishability should not increase when two states ρ and σ are subject to the same channel $\mathcal{N}$. This property is indeed satisfied by the fidelity and the trace distance and we thus have

$$F(\rho, \sigma) \geq F(\mathcal{N}(\rho), \mathcal{N}(\sigma)), \tag{2.90}$$

$$D(\rho, \sigma) \leq D(\mathcal{N}(\rho), \mathcal{N}(\sigma)). \tag{2.91}$$

2.4.3 Quantum Entropy

The entropy of a state of a physical system reflects the randomness of the system itself. A fundamental recent advancement in quantum mechanics represents the interpretation of quantum states as pieces of information. The concept of entropy was first expressed for thermodynamic systems of particles by Boltzmann between 1872 and 1875, for equally probable states, and generalized by Gibbs at the end of nineteenth century for not equally probable configurations of the system. Later, Shannon expressed the same concept in the field of information theory in 1948 which was actually inspired by von Neumann's generalization of Gibbs' definition of quantum-mechanical systems.

Figure 2.9 compares the three definitions of entropy. It shows the common structure of definitions of entropy in classical and quantum-physical systems: probability distributions are replaced by density operators and averaging over the phase space becomes a trace. Namely, while for a probability distribution $p(x)$, and any random variable X distributed according to p, the entropy is defined as

$$H(X) = H(p) := - \sum p(x) \log p(x).$$ (2.92)

For a quantum state ρ_X of a system, $\mathcal{H}_X$ is defined as

$$S(X) = S(\rho) = - \text{Tr}\left(\rho \log \rho\right).$$ (2.93)

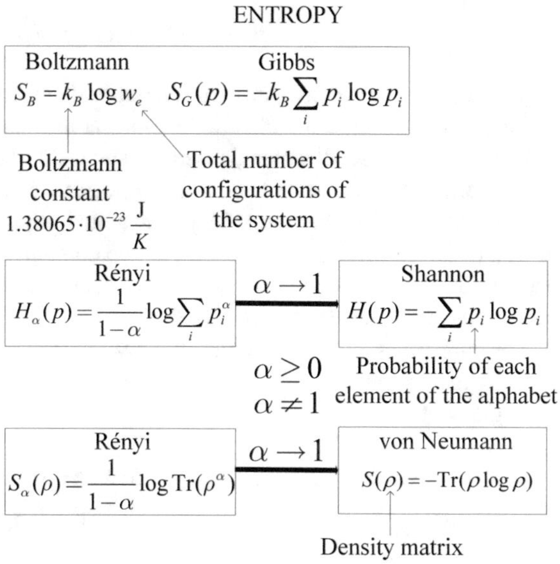

Fig. 2.9 Different expressions of entropy in thermodynamics (Boltzmann and its Gibbs generalization), information theory (Shannon), and quantum mechanics (von Neumann). Rényi entropy is a generalization of Shannon's and von Neumann's entropies; they are special cases for $\alpha \to 1$

ENTROPY

Boltzmann
$$S_B = k_B \log w_e$$

Gibbs
$$S_G(p) = -k_B \sum_i p_i \log p_i$$

Boltzmann constant $1.38065 \cdot 10^{-23} \frac{J}{K}$

Total number of configurations of the system

Rényi
$$H_\alpha(p) = \frac{1}{1-\alpha} \log \sum_i p_i^\alpha$$

$\alpha \to 1$

Shannon
$$H(p) = -\sum_i p_i \log p_i$$

$\alpha \geq 0$
$\alpha \neq 1$

Probability of each element of the alphabet

Rényi
$$S_\alpha(\rho) = \frac{1}{1-\alpha} \log \text{Tr}(\rho^\alpha)$$

$\alpha \to 1$

von Neumann
$$S(\rho) = -\text{Tr}(\rho \log \rho)$$

Density matrix

The definition of quantum entropy has been fundamental in looking at entanglement from the perspective of information theory. In information theory [Wil17], given two subsystems (random variables) X, Y, the joint Shannon entropy is

$$H(X, Y) \geq \max\{H(X), H(Y)\} \tag{2.94}$$

so the entropy of the whole composite classical system is always greater than those of single subsystems. On the other hand, given a quantum composite system with entangled subsystems, the quantum entropy of a subsystem can be greater than the joint quantum entropy

$$S(X, Y) \not\geq \max\{S(X), S(Y)\} \tag{2.95}$$

which means subsystems can have more disorder than the entire composite system.

What has just been mentioned is only the starting point of rethinking entropic inequalities: in fact, an information theoretic view of entanglement reveals significant differences between classical and quantum natures of communication. Because of the vast area of this research, in the context of this tutorial, another example is going to be shown.

Shannon's conditional entropy, defined as

$$H(X|Y) = H(X, Y) - H(Y) \tag{2.96}$$

is always positive. On the other hand, the quantum respective version

$$S(X|Y) = S(\rho_{XY}) - S(\rho_Y) \tag{2.97}$$

is guaranteed to be positive only for separable states. In fact, $S_{VN}(\rho_Y) \geq S_{VN}(\rho_{XY})$ can happen in the case of entangled non-separable states. *Can* has been voluntarily written because not all entangled states show this very interesting property and, indeed, any entangled state can be given positive conditional information with the addition of a local mixed state. Thus, the actual physical meaning of the negativity of $S_{VN}(X|Y)$ is still open for general mixed entangled states. An exhaustive survey and discussion of research on quantum entropic inequalities and their geometric perspective was presented in [HHHH09].

For pure states $|\Psi\rangle\langle\Psi|$ of a bipartite system $\mathcal{H}_X \otimes \mathcal{H}_Y$, the conditional entropy reduces to the entropy of the reduced states and it is known as the entropy of entanglement:

$$E(|\Psi\rangle\langle\Psi|_{XY}) = S(X) = S(Y) \tag{2.98}$$

$$= S(X|Y) = S(Y|X). \tag{2.99}$$

This measure is limited by the logarithm of the dimension of the Hilbert spaces

$$E(|\Psi\rangle\langle\Psi|_{XY}) \leq \log d_X, \log d_Y. \tag{2.100}$$

The pure states that saturate this bound are the *maximally entangled states*. For two-qubit maximally entangled states, sometimes referred to as *ebits*, this measure is one. The entropy of entanglement can be interpreted as measuring how many ebits are contained in the pure states, in the sense that asymptotically many copies of a bipartite pure state can be converted into maximally entangled states and vice versa, at a rate determined by the entropy of entanglement [BDSW96]. For mixed states, many inequivalent definitions of entanglement measures exist [PV07].

We have seen that the quantum conditional entropy displays a fundamentally different behavior with respect to the classical conditional entropy in the presence of entanglement. However, various important properties do find a generalization in the quantum setting. Most notably, the mutual information, and thus the conditional entropy, can still be written as a divergence, namely:

$$H(X : Y) = H(X) + H(Y) - H(XY) = D_{KL}(p_{XY} \| p_X \otimes p_Y) \tag{2.101}$$

$$S(X : Y) = S(X) + S(Y) - S(XY) = D(\rho_{XY} \| \rho_X \otimes \rho_Y) \tag{2.102}$$

where D_{KL} is the Kullback–Leibler divergence generalized in quantum states as the (Umegaki) relative entropy

$$D(\rho\|\sigma) = \begin{cases} \mathrm{Tr}\left[\rho(\log\rho - \log\sigma)\right] & \mathrm{supp}\rho \subseteq \mathrm{supp}\sigma \\ +\infty & \text{otherwise.} \end{cases} \tag{2.103}$$

The most important property that survives is the joint convexity. This is equivalent to a data processing inequality for relative entropy, also known as its monotonicity under quantum channels. Namely, for any states ρ and σ and any channel $\mathcal{N}$ it holds

$$D(\rho\|\sigma) \geq D(\mathcal{N}(\rho)\|\mathcal{N}(\sigma)). \tag{2.104}$$

Thus any inequality for the conditional information, mutual information, and further quantities (that is, derived as a consequence of the data processing inequality) still holds in the quantum setting. Most notably, while the conditional entropy can be negative, the conditional mutual information is always positive even in the quantum case of entangled states

$$H(X : Y|Z) = H(XZ) + H(YZ) - H(Z) - H(XYZ) \geq 0 \tag{2.105}$$

$$S(X : Y|Z) = S(XZ) + S(YZ) - S(Z) - S(XYZ) \geq 0 \tag{2.106}$$

for any tripartite state ρ of systems $\mathcal{H}_X \otimes \mathcal{H}_Y \otimes \mathcal{H}_Z$. The positivity of the conditional mutual information is actually equivalent to the data processing inequality and the joint convexity of the divergence ([Rus02] and references therein).

2.5   Bell Nonlocality

In the context of physics and relativity, the concept of *locality* consists of the absence of distant actions and the presence of strict bounds on the speed of a signal's transmission (i.e., speed of light). While quantum mechanics does not infringe on such a definition of locality, it shows some *non-classical* correlations (entanglement), which present *nonlocal* characteristics according to a classical definition of locality.

During the last century, theoretical and experimental physics proved that the predictions of quantum mechanics are incompatible with those of any classical physical theory satisfying a natural notion of reality while at the same time satisfying the locality principle imposed by special relativity. Namely, no deterministic *hidden variable* (HV) can model quantum-mechanical correlations of distant quantum systems. In that context, Einstein, Podolsky, and Rosen (EPR) proposed an experiment to show that quantum mechanics cannot be real (in the sense of describing reality by having observables being well defined at all times) and complete at the same time. Refusing to accept that a proper physical theory does not need to satisfy an axiom of reality, they used this to argue that quantum mechanics is incomplete. However, the correct resolution to their problem, now known as the EPR paradox, is to accept that distant observables can influence each other without being well defined and without violating locality at the same time.

A simplified version of the original EPR paradox can be described using qubits. Let us consider the state $|\Phi\rangle \sim |00\rangle + |11\rangle = |++\rangle + |--\rangle \in \mathcal{H}_A \otimes \mathcal{H}_B$ shared by Alice and Bob. If Alice performs a measurement with basis $\{|0\rangle, |1\rangle\}$, then the post-measurement state at Bob becomes $\{|0\rangle, |1\rangle\}$. Thus, Alice can predict the observable Z at Bob with certainty. Similarly, Alice knows the observable X at Bob with certainty if it measures $\{|+\rangle, |-\rangle\}$. EPR claimed that Z and X of System B's qubit were well defined independently of the measurement by a locality principle and this was in contradiction with the statement that Z and X cannot be well defined at the same time because of being non-commuting observables. Thus a problem of consistency or completeness must exist in quantum mechanics.

There are two types of locality, relativistically unified as a space-time locality. *Time locality* is the principle that real physical particles, fields, and generally all objects evolve depending only on their current state. This is reflected in the evolution in physics being described by differential equations. This means that physically, systems with memory do not exist. Models of systems with memory simply model hidden systems. Alongside this, *space locality* means that there is no action at a distance and all interactions are local with respect to space. By special relativity, the speed of light is universal, namely it is a fundamental speed limit for anything in the universe and can only be achieved by massless particles (these particles do not rest and they always travel at the speed of light even for observers with different speeds) as illustrated in Fig. 2.10. Nothing, even force, travels faster than massless particles, thus there cannot be a violation of Einstein's locality dictated by special relativity. Any field or force is inherently a function of the space-time position and it can only interact with the fields in this position.

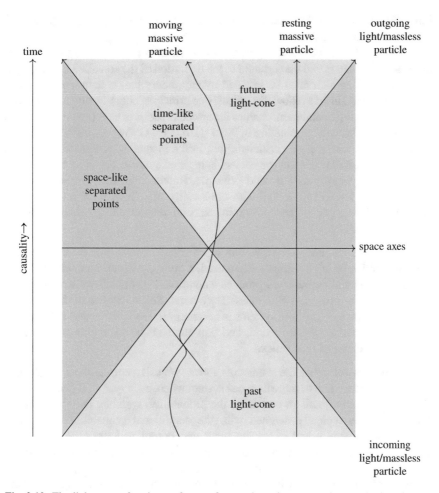

Fig. 2.10 The light-cone of a given reference frame where the space axis must be imagined as representing 3D space. Alternatively, this graph can be thought of as a 2D slice of 4D space-time. The red points are space-like separated from the origin of the reference frame and thus cannot influence or be influenced by the origin. The green points are the time-like separated points, divided in the past light-cone, the points that can influence the origin, and the future light-cone, the points that can be influenced by the origin. Lines in this graph represent the space-time path of some object and are called world-lines. Straight vertical world-lines represent the time evolution of resting particles, while world-lines parallel to the boundary of the cones are those of massless particles traveling at the speed of light. Massive particles can have curved world-lines but they are always constrained to be strictly contained in any light-cone originating in the line

The assumption of *local realism* claims that without disturbing a system, it is possible to predict with certainty (i.e., with probability equal to unity) the value of a physical quantity, so that there exists an *element of reality* corresponding to that quantity. On the other hand, quantum mechanics shows that two non-commutative observables like X and Z cannot have well-defined values simultaneously. Therefore, EPR argued that the reason for two non-commutative observables to be well defined simultaneously must lie outside of quantum mechanics.

In 1964, John Stewart Bell suggested an experiment that assumed the existence of hidden parameters and that brought to the definition of the so-called Bell inequalities [BCP$^+$14], which must be satisfied by any theory that is both real and local. This experiment aimed to show that at least one principle, either locality or realism is violated, assuming quantum mechanics is correct. A series of experiments starting in the 1980s [FC72] and culminating in 2015 [HBD$^+$15, GVW$^+$15, SMSC$^+$15] have shown, against the strongest criticism, that Bell's inequality is violated with high probability. Nowadays, many physicists agree that realism needs to be dropped, while keeping locality as the valid principle. Rather than involving actual time-space models, we will only deal with the locality of a finite number of systems. A simplified example of how to exploit special relativity to impose locality is illustrated in Fig. 2.11.

Remark 2.6 As we have seen, quantum mechanics is a statistical theory, meaning that irrespective of the quantum system used, a quantum experiment with classical inputs and classical outputs will produce a conditional probability distribution that might as well be simulated with classical systems. This is why involving multiple locality principles of multiple systems is necessary to see a distinction between the conditional probability distribution, and thus the statistical theories generated by quantum and classical mechanics.

We will now formalize the statements above. We call a correlation a conditional probability distribution of more than one system where each system is given an input (in the conditioning) and an output (in the probability) variable. For two systems, a correlation is typically denoted with the conditional probability distribution $p(a, b|x, y)$, where x and a are the input and output variable of Alice, and y and b the input/output variable of Bob. Locality of a finite number of systems is captured by the so-called *no-signaling correlations* or *non-signaling constraints*. These constraints physically imply that the local marginal probabilities of Alice are independent of either the input or output of Bob and vice versa: the outcome of experiments at Alice and Bob should not depend on each other. If the two systems are spatially separated, the no-signaling constraints impose that they cannot perform instantaneous signaling, thus avoiding the infringement of relativity mentioned above.

Formally, the local marginal conditional probabilities of each system for non-signaling correlations $p(a, b|x, y)$ must satisfy

$$p(a|x, y) = p(a|x) \quad \forall x \in X, a \in A, b \in B \tag{2.107}$$

$$p(b|x, y) = p(b|y) \quad \forall y \in Y, a \in A, b \in B. \tag{2.108}$$

More explicitly, the equalities are

$$p(a|x, y) = p(x|x, y') \quad \forall x \in X, a \in A, y, y' \in Y \tag{2.109}$$

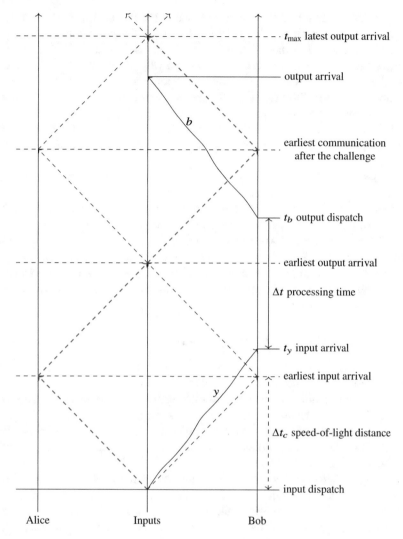

Fig. 2.11 A simplified version of Bell's experiment where Alice and Bob are placed at equal distances from the input source in opposite directions. The figure depicts how locality is imposed through timing restrictions using the speed-of-light principle. Bob is required to answer with the output before he has the chance to base his answer on information received by Alice, after Alice has received her input (red line). Because the speed of light is constant, distance can be measured in units of time by dividing any distance by the speed of light c. Δt_c is the distance of Alice and Bob from the inputs. The constraints imply that Bob has an interval of time of at most $2\Delta t_c$, the distance between Alice and Bob, to process the input and send the output

$$p(b|x, y) = p(y|x', y) \quad \forall y \in Y, x, x' \in A, b \in B, \tag{2.110}$$

where A, B, X, and Y are the alphabets which are assumed to be finite sets. If we denote the set of all correlations (with the above fixed input/output alphabets) as $\mathcal{C}$, then the set of all no-signaling correlations is denoted by $\mathcal{C}^{NS} \subset \mathcal{C}$. It is important to notice that both $\mathcal{C}$ and $\mathcal{C}^{NS}$ are polytopes[11] in $\mathbb{R}^{|X \times Y \times A \times B|}$ because since the alphabets are finite, the defining conditions are a finite number of all linear equations and inequalities (for the polytope of correlations these are the positivity of all elements summing to one, and for the non-signaling polytope the increasing number of linear equalities implies a smaller polytope). The vertices of the no-signaling polytope are deterministic states or non-deterministic Popescu–Rohrlich (PR) boxes.

The set of *local-hidden-variable correlations*, or *LHV* set, is the set of correlations compatible with a description via hidden variables shared by the parties, and thus represent the set of correlations being consistent with local realism. Bell argued that if physics obeyed local realism, then when measuring, for example, the spin of a given particle at Alice on direction **a**, the result can only depend on the hidden variable λ and on **a**, and similarly for the second particle at Bob. Then the distribution of the result is determined by the expectation over λ of the distribution of the two systems, which conditioned on λ, must be a product. Namely, an LHV correlation is any correlation of the form

$$p(a, b|x, y) = \sum_{\lambda} p_A(a|x, \lambda) p_B(b|y, \lambda) p_H(\lambda). \tag{2.111}$$

This property is also known as *Bell locality*. Notice that in this case the definition of the LHV correlation is already that of a correlation that can be written as convex combinations of deterministic correlations of the form $p_A(a|x) p_B(b|y)$. The set of all LHV correlations is denoted by $\mathcal{C}^{LHV}$, which is still a polytope, because there are only finite deterministic points. Notice also that all the extreme points of $\mathcal{C}^{LHV}$ are also non-signaling correlations and thus $\mathcal{C}^{LHV} \subseteq \mathcal{C}^{LHV}$.

Coming back to the example of the state $|\Phi\rangle$ shared by Alice and Bob, EPR used the fact that a measurement at Alice determines exactly the state at Bob for the two non-commuting observables X and Z to argue that a hidden variable must exist predetermining the outcome of X and Z, or otherwise the principle of locality would be violated. Since quantum mechanics does not predict this, it must either violate locality or be incomplete, but violation of locality would be out of the question, thus leaving only the conclusion that quantum mechanics is incomplete. However, the resolution is that the argument that the hidden variable must exist is wrong, and it is not required to not violate locality. The intuition built through our experience of the world obeys the LHV models and thus when prompted to think of non-signaling we immediately think of LHV rather than the correct condition. These

---

[11]A *polytope* is the convex hull of a finite set of points in some $\mathbb{R}^n$. The smallest convex set containing a point set $A \in \mathbb{R}^n$ is called the convex hull of A.

formal definitions show us that our intuition is already wrong, without bringing quantum mechanics into the picture.

To resolve the paradox, let us now analyze the *quantum correlations*, namely the correlations generated by quantum systems. Two quantum systems at Alice and Bob, $\mathcal{H}_A$ and $\mathcal{H}_B$, share a state ρ. By recalling the previous example with the spin-direction of the particles, the input classical variables $x \in X$ and $y \in Y$ at Alice and Bob are the variables determining which measurements to perform, and thus they each determine a POVM $\{M_{a|x}\}_{a \in A}$ and $\{N_{b|y}\}_{b \in B}$ respectively for each value.

Next, they measure the corresponding POVMs and they collect the outcome of the measurement. Thus, the distribution of outcomes, conditioned by x and y, are

$$p(a, b|x, y) = \mathrm{Tr}\left[\left(M_{a|x} \otimes N_{b|y}\right)\rho\right]. \tag{2.112}$$

By considering all possible collections of POVMs and all possible bipartite states, we construct the set of all quantum correlations denoted by $\mathcal{C}^Q$, which is still convex by the convexity of the quantum state, but generally is not a polytope (meaning it will have an infinite number of extreme points). The quantum correlations can be verified to be non-signaling by computing the margin of the correlation; the non-signaling conditions are then satisfied thanks to the defining property of the POVMs:

$$\sum_b p(a, b|x, y) = \mathrm{Tr}\left[\left(M_{a|x} \otimes \sum_b N_{b|y}\right)\rho\right] \tag{2.113}$$

$$= \mathrm{Tr}\left[\left(M_{a|x} \otimes \mathbb{1}\right)\rho\right], \tag{2.114}$$

$$\sum_a p(a, b|x, y) = \mathrm{Tr}\left[\sum_b \left(M_{a|x} \otimes N_{b|y}\right)\rho\right] \tag{2.115}$$

$$= \mathrm{Tr}\left[\left(\mathbb{1} \otimes N_{b|y}\right)\rho\right] \tag{2.116}$$

and therefore

$$\mathcal{C}^Q \subseteq \mathcal{C}^{NS}. \tag{2.117}$$

The effect displayed by quantum theory that led to the EPR paradox is the result that the set of quantum correlations is strictly larger than the set of LHV correlations

$$\mathcal{C}^{LHV} \subsetneq \mathcal{C}^Q. \tag{2.118}$$

Therefore quantum theory is local, but Bell nonlocal. The simplest proof that the maximally entangled state exhibits this behavior and produces correlations outside of the LHV set is presented below.

Remark 2.7 The set $\mathcal{C}^Q$ is produced by a physical theory, i.e., quantum mechanics. On the other hand, sets $\mathcal{C}^{NS}$ and $\mathcal{C}^{LHV}$ are defined as conditions or restrictions. It is common to associate $\mathcal{C}^{LHV}$ to the physical theory of classical mechanics;

however, a more accurate interpretation is to see it as the set containing any correlation produced by any physical theory that obeys locality and reality (that is, all observables always have a well-defined value).

To show the separation we need a general way of proving that a correlation is not an element of $\mathcal{C}^{LHV}$. This is done using Bell inequalities, which are a rephrasing of a general separation result from geometry.

Theorem 2.8 (Hyperplane Separation Theorem) *Let $P \subset \mathbb{R}^d$ be a closed convex set. Then there exists $S \in \mathbb{R}^d$ and $S_P \in \mathbb{R}$ such that*

$$S \cdot p \le S_P \quad \forall \, p \in P \tag{2.119}$$

but

$$S \cdot \hat{p} > S_P \quad \forall \, \hat{p} \notin P.$$

A Bell inequality is none other than a separating-hyperplane inequality for the set of LHV correlations. In this case the real space containing the hyperplanes is $\mathbb{R}^{|X \times Y \times A \times B|}$.

Definition 2.8 (Bell's Inequality) An inequality $S \cdot p \le S_{LHV}$ is called a Bell inequality if it is satisfied by any local-hidden-variable model p. An inequality representing a facet of $\mathcal{C}^{LHV}$ is called a tight Bell inequality.

Notice that since p is a conditional probability distribution, if $S \in \mathbb{R}^{|X \times Y \times A \times B|}$ is of the form $S_{abxy} = R_{ab}c_{xy}$, then the value $S \cdot p$ is actually a linear combination of expectation values of R

$$S \cdot p = \sum_{xy} c_{xy} \mathbb{E}[R|xy] =: \sum_{xy} c_{xy} \mathbb{E}[R_{xy}], \tag{2.120}$$

where R_{xy}, defined by linearity, is either a random variable or an observable depending on whether the conditional probability distribution is simply a correlation or if it is the result of a quantum measurement. Conversely, given any collection R_{xy} of random variables or observables (with a state ρ) we can construct an inequality by taking a linear combination of their expectation values. This gives a powerful template for constructing hyperplanes in the space of correlation by thinking only in terms of expectation values, whether classical or quantum. The most important case is the one where the expectation values are of local random variables or observables

$$C = \sum_{xy} c_{xy} \mathbb{E}[A_x B_y], \tag{2.121}$$

where in the case of quantum observables $A_x B_y \equiv A_x \otimes B_y$.

Among the Bell inequalities constructed this way, the *Clauser–Horne–Shimony–Holt* (CHSH) inequality considers a correlation $p(a, b|x, y)$ with $x, y \in \{0, 1\}$ and

$a, b \in \{+1, -1\}$, and is defined as

$$\text{CHSH} := \sum (-1)^{xy} \mathbb{E}\left[A_x B_y\right] \leq 2, \tag{2.122}$$

where $CHSH$ is called the *CHSH value*. More explicitly, it is possible to write the CHSH value as

$$\text{CHSH} = \mathbb{E}\left[A_0 B_0\right] + \mathbb{E}\left[A_0 B_1\right] + \mathbb{E}\left[A_1 B_0\right] - \mathbb{E}\left[A_1 B_1\right]. \tag{2.123}$$

Now, let us focus on the range of CHSH. For every p, $\mathbb{E}\left[A_i B_j\right] \in [-1, 1]$ since $a, b \in \{+1, -1\}$. This implies that $\text{CHSH} \in [-4, 4]$. This is achieved by the PR boxes and thus

$$\sup_{\wp^{NS}} \text{CHSH} = 4. \tag{2.124}$$

However, the value for $p \in \wp^{LHV}$ is indeed upper bounded by 2 as shown by

$$\sup_{\wp^{LHV}} \text{CHSH} \leq \sup \mathbb{E}\left[A_0\right] (\mathbb{E}\left[B_0\right] + \mathbb{E}\left[B_1\right]) - \mathbb{E}\left[A_1\right] (\mathbb{E}\left[B_0\right] - \mathbb{E}\left[B_1\right])$$

$$\tag{2.125}$$

$$\leq \sup a_0 (b_0 + b_1) - a_1 (b_0 - b_1) \tag{2.126}$$

$$\leq 2. \tag{2.127}$$

This upper bound can be achieved when Alice and Bob always reply $+1$. That is, $p(+1, +1 | x, y) = 1$. Thus, $\sup_{p \in \wp^{LHV}} \text{CHSH} = 2$.

Somewhere in between the NS and the LHV maximal values lies the maximal quantum value. Since we have upper bounded all the possible LHV correlations, to show that the quantum correlations can reach values higher than 2 we only need to provide a specific strategy that does it. We will let the parties share the two-qubit maximally entangled state

$$|\Psi\rangle = \frac{1}{\sqrt{2}}(|00\rangle - |11\rangle). \tag{2.128}$$

For the measurement, we considered the following rotated qubit bases in $\mathbb{C}^2$:

$$|0, \theta\rangle = \cos(\theta)|0\rangle + \sin(\theta)|1\rangle$$
$$|1, \theta\rangle = -\sin(\theta)|0\rangle + \cos(\theta)|1\rangle. \tag{2.129}$$

The standard basis is then the special case $|i\rangle = |i, 0\rangle$. For each input x and y Alice and Bob will perform a von Neumann qubit measurement in one of these bases. Therefore the POVMs of Alice and Bob will have the form

$$\left\{ M_{0|x}, M_{1|x} \right\} = \left\{ |0, \theta_x^A \rangle\langle 0, \theta_x^A |, |1, \theta_x^A \rangle\langle 1, \theta_x^A | \right\} \tag{2.130}$$

$$\left\{ N_{0|y}, N_{1|y} \right\} = \left\{ |0, \theta_y^B \rangle\langle 0, \theta_y^B |, |1, \theta_y^B \rangle\langle 1, \theta_y^B | \right\}, \tag{2.131}$$

which are used to measure the observables

$$A_x = M_{0|x} - M_{1|x} = |0, \theta_x^A \rangle\langle 0, \theta_x^A | - |1, \theta_x^A \rangle\langle 1, \theta_x^A | \tag{2.132}$$

$$B_y = N_{0|y} - N_{1|y} = |0, \theta_y^B \rangle\langle 0, \theta_y^B | - |1, \theta_y^B \rangle\langle 1, \theta_y^B |, \tag{2.133}$$

(the equivalent of Z operators in the rotated bases) and determine the answers $\{a, b\}$ with probability

$$\langle \Psi | A_x \otimes B_y | \Psi \rangle. \tag{2.134}$$

The actual choice of angles used to violate the CHSH inequality is

$$
\begin{aligned}
M_{a|0} &= |a, 0\rangle\langle a, 0|, \\
M_{a|1} &= |a, \pi/4\rangle\langle a, \pi/4|, \\
N_{b|0} &= |b, \pi/8\rangle\langle b, \pi/8|, \\
N_{b|1} &= |b, -\pi/8\rangle\langle b, -\pi/8|.
\end{aligned}
\tag{2.135}
$$

The achieved CHSH value is

$$CHSH = 2\sqrt{2}. \tag{2.136}$$

This value is known as Tsirelson's bound and it is actually the best one achievable with local quantum strategies alone (for the case of two players with single binary inputs and outputs). However, since the NS maximal value is 4, it means that there exist local correlations that cannot be achieved using local quantum mechanics. This actually shows that there is room for new physical theories that make predictions beyond the ones made even by quantum mechanics, that nonetheless still satisfy the locality principle.

2.5.1 Nonlocal Games

Another perspective to look at Bell's experiment/inequalities is through the viewpoint of game theory. In fact, Bell's experiments trying to violate a Bell inequality as much as possible can be interpreted as two (or more) cooperating spatially distributed parties, trying to satisfy an input–output relation as much as possible. Moreover, we can think of the inputs as being provided by a referee party which collects the answers and determines whether the parties have won when they provide output consistent with such relations. The violated Bell inequalities, and thus the non-LHV correlations, represent winning strategies that can only be accomplished by exploiting entanglement or stronger non-signaling correlations.

A well-known example is the *CHSH game* based on the CHSH inequality. In quantum game theory, this is a specific game, which has become important in both quantum mechanics (Bell's analysis and inequalities [BA12]) and theory of quantum communication networks [Nö19, LALS20]. In particular, the second application of the CHSH game was demonstrated to be useful to model the performance of quantum multiple-access channels (see Sect. 6.5) and to evaluate the communication complexity of quantum communications, exploiting entanglement. The following will introduce the fundamental terminology and definitions from both classical and quantum game theory, in order to understand these kind of quantum games and their applications in quantum communication networks. The description will only focus on useful definitions to understand CHSH cooperative games. A reader interested in quantum non-cooperative games can find a good description in [EW00].

Generally speaking, *game theory* is the study of rational decision-making in situations of conflict. It is an area of applied mathematics, which has been applied in various disciplines such as ecology, biology, politics, economy, and psychology [vNMR44]. However, with the formulation and evolution of quantum mechanics, part of the research community focused its interest on generalizing *classical* game theory to the quantum-mechanical domain, by involving quantum probabilities instead of classical ones [EWL99].

Classically, a *game* Γ of n players can be defined as the totality of the rules which describe it. A *play* is every particular instance at which the game is played (from beginning to end). Conventionally, the game Γ is a sequence of *moves* or *strategies*, whose value v is known. The single moves are denoted by $\mathfrak{M}_1, \ldots, \mathfrak{M}_v$ and assumed to be chronologically ordered.

In classical games, two kinds of moves are allowed. *Personal moves* are choices made by a specific player, depending on his own decisions. *Chance moves* are choices depending on external things, according to a specific probability. Every personal move should identify what player's decision determines it. The rules of the game restrict communication, i.e., the exchange of information among players. Finally, a *payoff function* is a function $\mathfrak{F}$, which assigns a real number (award) to each player at the end of the game. The scope of the game consists of maximizing this payoff function.

Nonlocal games are cooperative games[12] of incomplete information [CHTW04]. The quantity of information, shared in a given coalition, results in different values of payoff function. In fact, players can share none, some, or all (complete infor-

---

[12]A *cooperative game* evaluates the collective payoffs of coalitions of players, applying joint moves/strategies. On the other hand, a *non-cooperative game* consists of individual players with their own moves/strategies and payoff functions, evaluated via Nash equilibria.

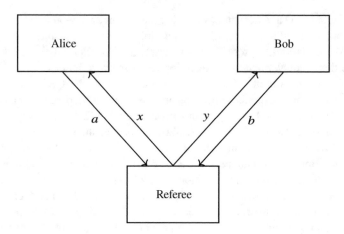

Fig. 2.12 Logic scheme of a quantum nonlocal refereed game (e.g., CHSH game), with two players (A and B) and a referee (R)

mation) of private information they own. Thus, cooperative games with incomplete information are also important to the concept of partition function.

The previously mentioned CHSH game is an example of a cooperative game with incomplete information. Figure 2.12 depicts the logic structure of such a game. The game consists of two cooperating players, A (Alice) and B (Bob), and a referee R (sometimes called a verifier). The referee starts the game, sending random questions to the players, drawn from finite sets, according to some chosen probability distribution. Next, players A and B are not allowed to communicate with each other, thus not knowing what has been sent to the other player. Players A and B answer to the referee, which finally evaluates players' answers according to his original questions. Through this procedure, he determines whether they win or lose. In a given nonlocal game, players have complete knowledge of the probability distribution that determines the referee's questions, as well as the condition, which decides whether they win or lose. Therefore, they can previously agree on their joint strategy.

Let $x \in \mathcal{X}$ and $y \in \mathcal{Y}$ be *questions* to Alice and Bob respectively. Next, let collections $\{A_{a|x}, a \in \mathcal{A}\}$ and $\{B_{b|y}, b \in \mathcal{B}\}$ be the descriptions of the measurements performed by Player A and Player B respectively. In the CHSH game, the referees send questions in the form of uniformly random bits $x, y \in \{0, 1\}$ and the players' goals are to produce outputs a and b, respectively, such that

$$a \oplus b = x \wedge y, \tag{2.137}$$

where $\wedge$ is the logical AND, which is 1 if all its arguments are 1 and 0 otherwise. However, they can also fail to satisfy this condition: in this case, they should maximize the probability of success.

Given this, the *classical value* of payoff function $\mathcal{F}_c$ of game Γ is the maximum probability with which Player A and Player B can win Γ, ranging over all purely classical strategies. The possible classical strategies are deterministic, private randomness, and shared randomness.[13] However, none of these classical approaches allows us to always satisfy condition Eq. (2.137) and thus, win the game. Even the deployment of shared or private randomness (instead of a deterministic solution) does not provide any advantage. In fact, the maximum classical success probability to satisfy Eq. (2.137) is $\mathcal{F}_c = 3/4 = 0.75 = 75\%$, which is the maximum achievable classical CHSH value: Alice and Bob always simply reply 0.

On the other hand, the only way to increase the value of the payoff function is to switch to quantum-mechanical strategies. Quantum games differ from classical ones mainly because they can employ quantum-mechanical phenomena and characteristics such as superposed initial states, superposition of strategies, and entanglement [GW07]. The entangled state and the local measurements, presented before for the CHSH inequality, also represent a quantum strategy for the CHSH game. The achieved CHSH value, shown previously, translates into the value of the payoff function of the CHSH game Γ (employing quantum strategies), so that it becomes $\mathcal{F}_q = \cos^2(\pi/8) = 0.853\ldots \approx 85\%$. This demonstrates that using quantum strategies is the only way to maximize the payoff function of the CHSH game beyond any classical strategy, even though is not the best non-signaling strategy.

2.6   Classical and Quantum Mechanics

Each mathematical or physical theory is based on some specific postulates. A *postulate* is simply a mathematical statement assumed to be true without proof. Physically however, postulates are meant to describe fundamental physical principles regarded as universal, which are assumed to be true until proven otherwise, usually justified by the supporting evidence. While classical and quantum theories have some key differences, they also share some common postulates:

- The evolution of the state of an isolated system at time t is only dependent on the state at time t. Mathematically, this translates into the evolution being described by differential equations. Classically, a differential equation will determine the

---

[13]When talking about classical communications, it is important to distinguish between deterministic (fixed) codes and randomized ones. The latter considers source and destination sharing randomness, which allows joint randomization of the encoderdecoder pair. While that does not affect the capacity, such shared randomness can improve the channel capacity in case the channel model is not deterministic (e.g., under adversarial communications, where a malicious node inserts jamming noise to disrupt communication). When shared randomness increases, the maximum achievable throughput equals the randomized coding capacity. Thus, the difference between randomized and deterministic coding capacity measures the maximum price to pay on the transmission rate, when communication parties do not share *classical* randomness [DJLS13, BBJ19].

evolution in phase space, while quantum-mechanically the Schrödinger equation determines the unitary evolution of the state in the Schrödinger picture, while the Lie bracket determines the same unitary evolution but in the Heisenberg picture.

- The state of the system determines the expectation value of an observable. An *observable* of a physical system is any physical variable of the system that can be measured, and for which a value can be assigned.

  The concept of separating the state from the observable is not usually encountered outside fundamental classical physics because of the implicit postulate concerning observables made by classical mechanics (below). Nonetheless, the idea is present in classical theory when considering, for example, lack of knowledge in the form of random variables (the observable) being distinct from the underlying probability distribution (the state).

That said, the remaining postulates differentiate quantum mechanics from classical mechanics. The postulates of classical mechanics are based on the intuition built on our perception of the world, which emerges as a limit of quantum mechanics in large scale. The postulates of quantum mechanics defy this intuition, but they are supported by the extremely accurate confirmation of experimental predictions.

For classical mechanics, some important postulates are

- *Determinism*: the state of the system is an observable and can be measured with arbitrary precision. All observables of a system are a function of the state and can also be perfectly measured (in principle), thus all observables have determined values and errors of measurement that can always be reduced below an arbitrary small threshold.
- *Separability*: given a set of physical systems, the state of the system can be determined via separate observables on each subsystem. Any uncertainty or measurement error is only due to lack of knowledge about the state of the system.

Due to the evident experimental violation of postulates of classical mechanics, the scientific community proposed different postulates for quantum mechanics such as

- *Superposition*: for a quantum system any linear combination of its states is also a valid state. The set of states is thus a vector space.
- *Linear observables*: Observables are described by the space of linear operators (functions) on the vector space of states that have a real spectrum. The spectrum are the values that the observable can assume. The space of observable is thus a non-commutative space. An observable can be said to have a certain determined value only on eigenvector states, also called eigenstates.
- *Expectation values*: the expectation value of any observable is the inner product defined by the observable on the given state. The set of all expectation values determines the state of an isolated system. Because an observable does not have

a determined value on every state, expectation values of observables are the only measurable quantities in quantum mechanics.[14]

- These postulates in particular imply

  - *Non-separability*: given a set of quantum systems, there exist states (the entangled states) that are not determined as the independent states of the subsystems, and all their physical properties cannot be separately determined as the result of independent measurements.
  - *Complementarity*: there exist observables that do not commute, these are observables for which the order of the measurement matters and return different expectation values. Interference is a particular case of the behavior induced by the eigenstates of one observable on a different non-commuting one.
  - *Statistical theory*: because quantum theory separates states from observables and defines the expectation value as it is, quantum theory is intrinsically a statistical theory, and the outcome values of experiments can be determined only probabilistically.

The validity of quantum theory is supported by its ability to give accurate predictions in experiments that reveal quantum-mechanical behaviors like photoelectric and Compton effects, black body radiation, specific heat, electron diffraction, structure of atoms, etc.

Remark 2.8 While the postulates of quantum mechanics might seem like restrictions compared to classical mechanics, where any function of the state can be an observable, this is not the case. Classical theory is recovered in the spectrum of commuting observables measured on their eigenstates.

---

[14]Such a statement considers measurable quantities that can be reproduced only. Consider an experiment that delivers outcomes in a certain set. We can think of the experiment itself as preparing a state and measuring an observable X with a given spectrum. If the experiment measures a non-eigenstate and outputs a certain value x, we cannot say to have measured X, because performing the exact same actions of the experiment again will return different outcomes. Thus a single outcome cannot be considered a meaningful measurement by itself. The only reproducible measurement is the expectation value estimated over many runs of the experiment, and this value will indeed not change (within precision) if we repeat the estimation process.

Chapter 3
Quantum Computing and Programming

The reason behind developing quantum computing is not only related to achieving higher computing capacity but also to be able to properly simulate and numerically study quantum systems. In fact, simulating quantum systems explicitly require exponentially many degrees of freedom, and indeed, classical computers cannot efficiently simulate all the possible dynamics [BA12] (for some restricted classes, the explicit simulation can be avoided, and the dynamic can be simulated efficiently in the system size). It is still not clear what all the computational tasks for quantum networking are and which of them need to be or are preferably solved with a quantum computer. The goal of this chapter is to provide an understanding of these key quantum ideas that are standard must-do topics in quantum computation and which might find their applications in quantum networks. Aside from this, the general hope is that since the intuition behind quantum algorithms is so different from classical ones, thinking of algorithms quantum-mechanically will eventually lead to new developments and ideas in quantum computation that cannot be achieved with classical thinking.

Part of the success of classical computers is due to the possibility of computing (calculating) anything using a small set of instructions. More precisely, while the whole of the physics in the universe is not known, the physics that is known is such that anything that can be produced can be simulated or computed using a small set of operations given enough time and information space. This has led to the belief that any function that can be computed by any means can also be computed by a suitably large digital computer; this is also known as the Church–Turing thesis, a conjecture about the physical world. Focusing on few instructions has allowed classical computers to develop, improve, and optimize, irrespective of their purpose or application. The process of decomposing an operation into sequences of instructions in a set is known as compilation.

© The Editor(s) (if applicable) and The Author(s), under exclusive license to
Springer Nature Switzerland AG 2021
R. Bassoli et al., *Quantum Communication Networks*, Foundations in Signal
Processing, Communications and Networking 23,
https://doi.org/10.1007/978-3-030-62938-0_3

3.1   Universal Gate Sets

Basic instructions that are not decomposed further into other instructions are called gates. A set of gates that allows us to compute any function is called a universal gate set, and the task of implementing a given function using such a set is called a computational problem. It is known that Boolean gates (NOT, AND, NAND, OR, NOR, XOR, etc.) can be used to compute any finite function, also known as functional completeness, and they are building blocks for digital circuits. This in particular means that the set of Boolean gates constitutes a universal gate set. Moreover, all the Boolean gates together are redundant for universality, and thus different universal gate sets can be constructed with different Boolean gates, each set leading to a different model of computation.[1] Known examples are { AND, NOT }, { NOR }, and { NOR }. The expression *A gate is/is not universal* is commonly used to convey whether this gate, as a set of one element, constitutes a universal gate set, and not whether it is part of a universal gate set (once a set is universal, adding more gates trivially creates a new universal gate set, and therefore the latter interpretation is not really meaningful).

The same is true for quantum computation. All possible quantum operations can be approximated arbitrarily well using a small set of fixed quantum operations [Kit97a], and thus it is possible to find a universal quantum-gate set defining a model of quantum computation.[2] Different models of computation also exist for quantum computation. As of this writing, no fully working quantum processor exists and no physical architecture has fully supplanted the others. We are thus still some milestones away from fixing and adopting a standardized set of quantum gates.[3] The following is one of the most common ways of constructing a universal quantum-gate set and computation model.

3.1.1   Quantum Circuit Model

The circuit model of quantum computation provides a graphical representation of quantum algorithms. In the strict version, only unitaries acting on pure states are represented, with an optional measurement at the end. Qubits are organized in

---

[1]Classical computers have evolved so far as to even standardize the higher level instruction sets offered by commercial processors (x86, ARM, etc.).

[2]As opposed to classical operations, quantum operations cannot be implemented exactly with a finite or countable number of gates due to the set of operations on any number of qubits being continuous. The only thing we can hope for is to generate a dense subset, just like natural numbers generate the dense set of rational numbers in the real numbers. Even if the exact implementation was possible, simple approximation is the physically relevant condition, since due to the impossibility of performing exact operations and measurements in practice, the outcome of a real-world quantum computation can only be obtained with a small but non-zero probability of error.

[3]Let alone a standardized set of quantum processor instructions.

Table 3.1 Some single-qubit gates and examples of their applications

Qubit gate	Circuit	Matrix
Hadamard	H	$H = \frac{1}{\sqrt{2}} \begin{bmatrix} 1 & 1 \\ 1 & -1 \end{bmatrix}$ $\|a\rangle \to \frac{1}{\sqrt{2}}(\|0\rangle + (-1)^a \|1\rangle)$
Pauli X (Bit flip, NOT)	X	$X = \begin{bmatrix} 0 & 1 \\ 1 & 0 \end{bmatrix}$ $\|a\rangle \to \|a \oplus 1\rangle$
Pauli Y (Bit&Phase flip)	Y	$Y = \begin{bmatrix} 0 & -i \\ i & 0 \end{bmatrix}$ $\|a\rangle \to i(-1)^a \|a \oplus 1\rangle$
Pauli Z (Phase flip)	Z	$Z = \begin{bmatrix} 1 & 0 \\ 0 & -1 \end{bmatrix}$ $\|a\rangle \to (-1)^a \|a\rangle$
Phase gate (S or P gate)	S	$S = \begin{bmatrix} 1 & 0 \\ 0 & i \end{bmatrix}$ $\|a\rangle \to (-1)^a \|a\rangle$
Phase shift/rotation	$R(\alpha)$	$R(\alpha) = \begin{bmatrix} 1 & 0 \\ 0 & e^{i\alpha} \end{bmatrix}$ $\|a\rangle \to e^{ia\alpha} \|a\rangle$
Z Measurement		Not a matrix $C \otimes \|a\rangle \to CC^\dagger \otimes \|a\rangle\langle a\|$
Serial gates	A_1 — A_2	$A_2 A_1$

parallel single lines, where each line represents the time steps of a single qubit. Gates can act on one or multiple qubits; Tables 3.1 and 3.2 define and show the most important gates to know. The output of a measurement is not a pure state unless conditioned on the measurement outcome; however, because all measurements in a protocol/operation can always be moved to the end using purification procedures, measurements are allowed as the last operation on a qubit (see Table 3.1). Double lines are used to represent classical values, and in particular, the outcome of the measurements, which can be used to condition further operations on other qubits. After a measurement, which includes tracing or discarding, a qubit is considered to be "destroyed" (practically, the physical qubit will be ready for reinitialization and further use, only the information content is destroyed).

Optionally, all qubits might be required to be initialized to $\|0\rangle$, as some computation might be hidden in the input quantum state. For example, the input

Table 3.2 Some two- and three-qubit gates

Multiqubit gate	Circuit	Matrix
CU (Controlled U)		$\begin{bmatrix} 1 & 0 & 0 & 0 \\ 0 & 1 & 0 & 0 \\ 0 & 0 & & \\ 0 & 0 & & U \end{bmatrix}$ $\lvert a,b\rangle \rightarrow \lvert a\rangle \otimes U^a \lvert b\rangle$
CNOT (Controlled X)		$\begin{bmatrix} 1 & 0 & 0 & 0 \\ 0 & 1 & 0 & 0 \\ 0 & 0 & 0 & 1 \\ 0 & 0 & 1 & 0 \end{bmatrix}$ $\lvert a,b\rangle \rightarrow \lvert a, a\oplus b\rangle$
Z (Controlled Z)		$\begin{bmatrix} 1 & 0 & 0 & 0 \\ 0 & 1 & 0 & 0 \\ 0 & 0 & 1 & 0 \\ 0 & 0 & 0 & -1 \end{bmatrix}$ $\lvert a,b\rangle \rightarrow (-1)^{a\cdot b}\lvert a,b\rangle$
SWAP		$\begin{bmatrix} 1 & 0 & 0 & 0 \\ 0 & 0 & 1 & 0 \\ 0 & 1 & 0 & 0 \\ 0 & 0 & 0 & 1 \end{bmatrix}$ $\lvert a,b\rangle \rightarrow \lvert b,a\rangle$
Toffoli		$\begin{bmatrix} 1 & 0 & 0 & 0 & 0 & 0 & 0 & 0 \\ 0 & 1 & 0 & 0 & 0 & 0 & 0 & 0 \\ 0 & 0 & 1 & 0 & 0 & 0 & 0 & 0 \\ 0 & 0 & 0 & 1 & 0 & 0 & 0 & 0 \\ 0 & 0 & 0 & 0 & 1 & 0 & 0 & 0 \\ 0 & 0 & 0 & 0 & 0 & 1 & 0 & 0 \\ 0 & 0 & 0 & 0 & 0 & 0 & 0 & 1 \\ 0 & 0 & 0 & 0 & 0 & 0 & 1 & 0 \end{bmatrix}$ $\lvert a,b,c\rangle \rightarrow \lvert a,b, ab\oplus c\rangle$

Table 3.3 Some multipartite entangled pure states

State	Circuit
Bell states $\|\phi_{xy}\rangle = \frac{1}{\sqrt{2}} \sum_i (-1)^x \|i, y \oplus i\rangle$	
GHZ states $\|n\text{-GHZ}\rangle = \frac{1}{\sqrt{2}}(\|0\rangle^{\otimes n} + \|1\rangle^{\otimes n})$	

computation might initially be classical and need encoding into a quantum state as the first step. Table 3.3 shows how to produce some of the important entangled states starting from initialized qubits. Additional initialized qubits can be used to aid or simplify the circuit of an operation. However, since discarding them might affect the computation due to entanglement, unlike classical bits, special care is needed to return them into a discardable state. These are called *ancilla* qubits. For example, in Table 3.2, Toffoli can be used to implement a CNOT, e.g., by setting $|a\rangle = |1\rangle$, which can then be safely discarded or returned to $|0\rangle$; in this case, $|a\rangle$ acts as an ancilla qubit.

Compared to classical circuits, some operations are prohibited. Since the columns of the circuit represent time steps, loops in such circuits are not allowed. Moreover, due to no-cloning, fan-in and fan-out of quantum systems are also not possible, since the information of one qubit cannot be copied into two qubits, and erasing a qubit is a noisy operation. Circuit-like diagrams are also used in quantum information to represent operations in density matrix formalism. These lines are not restricted to represent single qubits, and elements like fan-in and fan-out of quantum operations simply represent dealing with multiple systems. Tables 3.1 and 3.2 present some of the most important quantum gates used in the circuit model.

By focusing on the physical realization of quantum processors and quantum information processing, in 2000 DiVincenzo published a seminal article [DiV00], describing some important guidelines toward actual quantum computing. These are also known as the seven *DiVincenzo's criteria*:

1. *A scalable physical system with well-characterized qubits.* The quantum computing system should have the capability to contain well-characterized qubits, which are qubits whose physical parameters are accurately known, including their internal Hamiltonian. Moreover, it should be known if there are couplings among states and interactions with other qubits and/or external fields.

2. *The ability to initialize the state of the qubits to a simple fiducial state.* A pillar of classical computing is that registers can be initialized to a known value before starting processing of data. Moreover, quantum error correction needs a continuous supply of qubits in a low-entropy state (see Chap. 5).

3. *Long relevant decoherence times, much longer than the gate operation time.* As previously mentioned in Sect. 2.3, each quantum system—and so a qubit—gets coupled with either the equipment or the environment, causing decoherence. For correct quantum computing, the decoherence time must be long enough not to have impact on computational operations. This can be determined using results of quantum error correction theory (see Chap. 5). However, it is important to keep in mind that decoherence is a system-specific phenomenon, so it depends on the details of all the specific realizations of qubits couplings to various environmental degrees of freedom.

4. *A universal set of quantum gates.* An algorithm for quantum computing is specified as a sequence of unitary transformations, processing a small number of qubits. Physically, this implies the characterization of the Hamiltonians that produce such unitary transformations. The main objective is to have quantum computing that does not entangle with its control apparatus (in reality the entanglement exists but it should be negligible). In actual realizations, quantum gates are not perfect. Thus, they experience systematic and random errors in the implementation of the Hamiltonians. That represents an additional component of decoherence that requires error correction techniques.

5. *A qubit-specific measurement capability.* In Sect. 2.3, the centrality of the measurement of quantum systems was explained. A key point of quantum computing—and computing in general—is the capability of reading the results of computations. Unfortunately, the efficiency of real quantum measurement is not 100%, but this is not a fundamental issue since less can be required by quantum computations. In fact, what becomes important is the trade-off between quantum efficiency and reliable computations. What lower values of reliability imply is the parallel increase of the repetition of calculations. Alongside the trade-off between complexity and reliability, there is also the one between measurement time and decoherence time. In particular, faster measurements permit more time for error correction procedures.

6. *The ability to interconvert stationary and flying qubits.* In quantum communication networks, there are two kinds of qubits: the ones used *locally* for quantum computations (stationary) and the other ones that are really transmitted between different places (flying qubits). Practical examples of the latter may be qubits encoded in the polarization of photons—transmitted via optical fibers, or the usage of electrons, traveling in solids.

7. *The ability to faithfully transmit flying qubits between specified locations.* This requisite is closely interconnected to the previous one, which represents a complement. While the previous requisite focuses on the interface between communication and computing of information, this seventh purely considers the optimal creation and detection of flying qubits.

Tensor Networks

With quantum circuits, quantum-mechanical systems are described only as vector/pure states and operations with unitaries that eventually end in a measurement. Crudely speaking, the reason why this works is that there is a one-to-one correspondence between the lines in and out of one of these objects and the type of object in terms of the number of indices of the represented tensor, which can be called *legs*. Connected lines represent contractions of indices/legs, namely indices of different tensors that are summed over, while open lines mean open indices, i.e., open to matrix multiplications.

More precisely, a complex number has no indices and thus has no legs. A ket vector has one right facing leg to start the circuit, and bras multiply kets to produce complex numbers. Thus they have left facing legs that can be connected to right facing legs. Linear operators, being linear combinations of kets and bras, have thus a left and a right facing leg. Objects next to each other without contracted legs are in tensor product. It should now be clear that it is possible to generalize the diagrams to arbitrary tensors, each leg representing a ket or a bra entry in the tensor itself.

However, lines can be used to represent any kind of vector index, even if the bra–ket notation is not used for them. The vertical line of the CNOT, for example, can also be thought of as a contracted index. Since the CNOT is the sum of $|k\rangle\langle k|$ in tensor product with X^k, both $|k\rangle\langle k|$ and X^k can be thought of as tensors with three legs contracted over the leg indexed by k. Even more, connected legs coincide with noiseless quantum channels, and therefore the size of the index in a connection represents the potential amount of quantum information equivalent needed to implement the contraction. In the case of the CNOT, the size of the index of the internal and external legs is equal, which reflects the fact that if we want to make a CNOT between distant qudits, a qudit of communication must be performed to bring them together.

The deep study of the characteristics of entanglement and quantum many-body systems requires complex mathematical formulations. However, it is often reasonable to assume that full quantum channels between distant particles are suppressed by chaos. Therefore a reasonable ansatz on the state of the many bodies is that it is generated by many contracted tensors where the connectivity represents an assumption on the amount of entanglement that naturally propagates through the system. This turns out to be an important mathematical representation that has been growing in importance in the description of entanglement and complex quantum systems.

Tensor networks [WBC15, Orú19] are important mathematical tools for graphic calculus, which have the aim of translating problems into graphical structures. The first graphical notation to represent quantum states, operators, and maps was proposed by Penrose. Figure 3.1 depicts the respective graphical representation of Dirac's algebraic notation and the notation based on operators for quantum mechanics. A thorough introduction to the topic can be found in [Orú14].

By drawing quantum states with networks of interconnected tensors, the structure of entanglement among the constituents of the system appears clearer. Moreover,

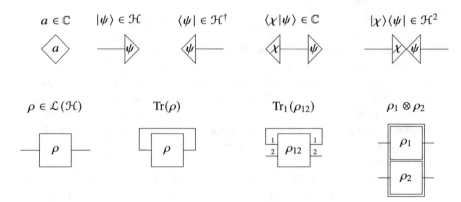

Fig. 3.1 Notation of tensor networks and correspondence with Dirac's notation

this structure can be different according to the dimension of the composite system (1d, 2d, etc.). As previously mentioned, entanglement is a quantum correlation independent of distance, and tensor network diagrams very effectively reflect entanglement in their graphical correlations.

3.1.2 Quantum Universal Gate Sets

In the classical circuit model of computation, the sets of classical gates { NAND }, { AND, NOT, XOR }, and { Toffoli }, are universal. Referring to Tables 3.1 and 3.2 for the definitions, in the quantum circuit model we can say the following statements about quantum gates:

- Sets of single-qubit gates cannot be universal (also true classically, but there the only single-bit gate is the NOT), in particular the Pauli group $\mathcal{P}_n$ generated by $\{ X, Y, Z \}^{\otimes n}$ cannot be universal.
- Sets composed of real gates can only produce real gates and thus cannot be universal.
- Sets that can approximate any real gate arbitrarily well can also *simulate* universal gate sets with the use of an ancilla qubit [RG02]. Part of the trick is to split the real and imaginary parts as $|\psi\rangle \to |0\rangle \otimes \Re(|\psi\rangle) + |1\rangle \otimes \Im(|\psi\rangle)$ for states and as $U \to \begin{pmatrix} \Re(U) & \Im(U) \\ -\Im(U) & \Re(U) \end{pmatrix}$ for operators. These gate sets cannot, however, achieve any unitary operator in the space of all the qubits they act on (ancilla included).
- It is enough to pair Toffoli with a single-qubit real gate that does not preserve the standard basis, like Hadamard, to be universal [Shi03]; in particular, { Toffoli, H } is universal [Kit97a]. Similarly, it is enough to pair CNOT with a single-qubit real gate such that the square does not preserve the standard basis [Shi03].
- { CNOT, H, $R(\frac{\pi}{8})$ } is universal [BMP+00].

- { H, S, CNOT } generates what is called the Clifford group, it contains the Pauli group and is defined by the property that any element of the Clifford group maps Pauli gates into Pauli gates under conjugation. It was proven that circuits made of Clifford gates can be simulated efficiently [Got97, Got98a, AG04], and are thus not universal.

Clifford gates have a special role in the construction of universal gate sets. As we will see, error correction codes are mostly based on Pauli measurements with the effect of collapsing any error into Pauli errors. Furthermore, the correction of detected errors itself might introduce new errors, and thus a quantum computer should try to collect the correction of errors detected at different times whenever possible. By their property, Clifford gates can commute with the detected errors with the effect of transforming the Pauli error into a different Pauli error only. This can be tracked efficiently with classical computation, thus allowing us to defer the correction of all the errors in a sequence of Clifford gates to the end of the sequence. This interplay makes non-Clifford gates more costly and makes it important to reduce their number, even at the cost of increasing the number of ancilla qubits or Clifford gates (just like multiplications are considered costly compared to additions in classical computation). { H, S, CNOT } are thus usually included in any universal gate set even if some gates are redundant, this is the first step in reducing the number of non-Clifford gates and forced-triggered correction operations.

In conclusion, the problem of compiling even a few-qubit gates is far from straightforward. A lot of work is focused exploring the trade-off and optimal decomposition of single-qubit gates. In order to keep the exposition simple in the algorithms to come, but somehow keep them overall complete, we will assume that all single-qubit gates can be implemented efficiently, and only make the following remarks; for further reading, see [BBC+95, KMM13, RS16, BGS13].

- Just like light polarization, single-qubit gates can be decomposed in different ways into at most five rotations among $e^{i\theta_x X}$, $e^{i\theta_y Y}$, and $e^{i\theta_z Z} = R(\theta_z)$. Rotation gates are usually easy to implement in physical qubit implementations, where often the parameter θ directly relates to the time of a driving signal. Therefore rotations, or all single-qubit gates, are sometimes considered part of the universal gate set. However, they might not be so natural in logical qubits, namely, as the gates of an already error corrected qubit.
- With a discrete universal gate set, any single-qubit unitary can be approximated with a number of gates that scale as $\log(1/\varepsilon)$, where ε is the precision of the approximation under a suitably defined distance measure.

3.2 Computational Complexity

The importance of gates (and instruction sets in general), and the reason why they define different models of computation, is that such sets are usually associated with unit cost of operations. Namely, adopting a given gate set assumes that each gate in the set has a unit cost, and the cost of any other operation is the number of gates necessary to compile it. Clearly, given an operation, its cost under different gate sets will be different. Furthermore, if the cost is time, comparing the same operation on different machines will result in different times that change by a multiplicative factor, since the time unit spent in the gates is different. This, together with the fact that the input length to a task is usually variable (the typical example is compression), makes computational complexity the relevant measure for comparing the efficiency of algorithms.

Given a problem family, defined for any input length n and an algorithm that provides the solution, the computational complexity of the algorithm is defined as the scaling of the cost of the algorithm as a function of n. In the field of computational complexity, the relevant measure is usually the worst-case complexity, namely, the scaling as a function of n of the largest cost over all inputs; however, for practical purposes, the average-case complexity is also relevant, where for fixed n the cost is computed by the average cost over all inputs. The complexity of an algorithm is an intrinsic property and does not change if the machine that runs it changes, except when switching from quantum to classical, if the universal gate set is changed.

An algorithm is said to be *efficient* if its complexity is polynomial in n, in which case it is said to be a polynomial-time algorithm or to run in polynomial time. It is always important to remember that the complexity scaling is the important measure *in the limit* of large inputs, and computational efficiency does not translate directly into practical efficiency.[4]

The complexity of a problem itself is the lowest complexity of all algorithms that provide a solution, and a problem is said to be efficiently solvable if there exists an efficient algorithm that provides a solution, otherwise it is said to be a hard problem. For some problems, we know of efficient quantum algorithms, but we do not know if any efficient classical one exists.[5] To simulate a quantum algorithm with a classical one, the pure state of the qubits must somehow be simulated. While in an efficient quantum algorithm the operations on the pure state of n qubits are naturally done using polynomially many quantum gates, the classical simulation might have to store the pure state explicitly, which means storing $\sim 2^n$ vector coefficients, making the simulation inefficient as soon as one computation on the whole vector is performed. Therefore, there could exist problems that are simply

---

[4]Some of the reasons for this are the following. Algorithms are still time-consuming if the complexity is a polynomial of high degree. Different cost factors come into play in practice, where some basic operations can be much more expensive than others (multiplication vs. addition) and memory access must be taken into account (cache vs. RAM operations). Sometimes the hidden constants can be so large to make the scaling advantage impractical and the algorithm unusable. For example, multiplication of $n \times n$ matrices must use at least n^2 scalar multiplications (the number of output values), and implementing the definition without optimization uses n^3 multiplications. Matrix multiplication was proven to have a complexity of at most $n^{2.4}$[CW90]; however, the algorithm that was found is impractical and yields no advantage in today's computers[Ili89]. A practical algorithm exists that uses $n^{\sim 2.7}$ multiplications [Str69].

[5]All such examples are related to the complexity of the Fourier transform, which we will see later.

quantum-mechanically but not classically efficient, namely that can only be solved efficiently with a quantum computer. Such a question is still unanswered and is related to other fundamental long standing problems in complexity theory, the field studying the classification of problem complexity.

Finally, it should be clear that since classical mechanics is a special case of quantum mechanics, quantum computation can always compute just as fast as classical computation. This can be readily seen by writing any classical computation in density matrix formalism, but it is also true even under the restriction of using a computation model like the one above, namely if we are forced to use pure states and unitary operations with measurements only at the end. This is due to the possibility of efficiently converting any classical algorithm into a reversible classical algorithm using only classical Toffoli gates via, for example, Bennett's uncomputing trick [Ben73, BBBV97]. All Toffoli gates can then be made quantum, and the algorithm can be implemented with the addition of measured ancilla qubits.

Quantum speedup refers to any situation where the complexity of a quantum algorithm is lower than the best known classical algorithm. The cases mentioned above, where the speedup is such that the quantum algorithm is efficient while the best classical algorithm is not, are examples of exponential speedups, and such quantum algorithms are those that lead or can lead to breaking cryptographic functions. Many quantum speedups are formulated in terms of oracles, either because the problem itself is an oracle problem, or to simplify the exposition. We thus still need to introduce the notions of classical and quantum oracles before we can properly explain the most important quantum algorithms. Before we do this, in order to add concreteness to the concepts introduced so far though, we cover the compilation of the quantum Fourier transform.

Remark 3.1 To compare classical and quantum algorithms for speedup, they must solve the same problem.

We have mentioned that a single-qubit gate has a complexity scaling of $\log(1/\varepsilon)$, where ε is a distance. Under the assumption that the required precision does not depend on the size of the circuit (a reasonable assumption if the code is run on error corrected qubits as we will explain in more detail in the next section), this implies efficiency and constant complexity. Thus, for the sake of simplicity, we will take this for granted and not decompose further single-qubit gates.

3.3   The Quantum Fourier Transform

The Hadamard gate is a unitary transformation that changes the two basis states $|0\rangle$ and $|1\rangle$ of qubit $|\psi\rangle$ into an unbiased basis: this basis is also called the *Hadamard basis*, X basis, or conjugate basis for reasons that will be clear below. It is an example of the Discrete Fourier Transform (DFT) used as a quantum operation, known as the Quantum Fourier Transform (QFT), which is possible due to the DFT being a unitary matrix and thus a valid quantum evolution. We will make no distinction between discrete and quantum and simply call it the Fourier transform. For a qudit of dimension d, the Fourier transform on the computational basis is defined as

$$F := \frac{1}{\sqrt{d}} \sum_{xy \in \mathbb{Z}_d} \omega^{-xy} |x\rangle\langle y| \qquad \text{with } \omega := e^{i\frac{2\pi}{d}}. \qquad (3.1)$$

Recall that for qudits, X and Z are defined such that $X |x\rangle = |x + 1\rangle$ and $Z |x\rangle = \omega^x |x\rangle$. The sign choice in $\omega^{\pm ij}$ is such that $F |i\rangle$ is the conjugate eigenbasis, just like conjugate classical variables are pairs of variables related by a Fourier transform, namely such that $XF |x\rangle = \omega^x F |x\rangle = FZ |x\rangle$, or simply $XF = FZ$,

$$-\boxed{Z}-\boxed{F}- \quad = \quad -\boxed{F}-\boxed{X}- ,$$

which also reveals that X has the same spectrum as Z.

If n qubits are present, $H^{\otimes n}$ is *a Fourier transform* but does not implement *the Fourier transform*. More precisely, it is the Fourier transform defined by the binary sum of n bits rather than the integer sum modulo 2^n, namely, it satisfies $XH = HZ$ and thus $X^{\otimes n} H^{\otimes n} = H^{\otimes n} Z^{\otimes n}$ with X and Z the qubit Pauli operators rather than the 2^n dimensional relation $XH = HZ$ with the qudit X and Z operators. $H^{\otimes n}$ is clearly efficient, and in particular it has the same effect on $|0\rangle$ as the Fourier transform. Therefore $H^{\otimes n}$ can always be used to generate the uniform superposition state $\sum_x |x\rangle$ rather than using the more expensive 2^n dimensional Fourier transform F. The question is whether F is also efficient whenever $H^{\otimes n}$ cannot be used.

The trick for efficient computation of the Fourier transform F is to write x and y explicitly from their bitstrings via $x = x_{n-1} \ldots x_0 = \sum_{k \in \mathbb{Z}_n} 2^k x_k$, and the product xy will then expand into terms involving only two bits at a time that can be translated into Hadamard gates and controlled phases $CR(\alpha)$, a special case of controlled unitary from Table 3.2. The reader can find the details in, e.g., [NC10]; here, we simply show the circuit in Fig. 3.2 and explain how to implement the controlled phases. We define the rotations $R_n := R(2^{-n})$, in particular $R_0 = \mathbb{1}$, $R_1 = Z$, and $R_2 = S$; the controlled phases used in Fig. 3.2 are simply CR_n. Any controlled phase can be implemented as [KC18]

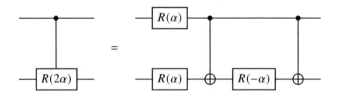

which reduces to the compilation of single-qubit gates that we said is efficient.

The number of gates in the circuit is of order n^2, with the compilation of the controlled phases only costing a multiplicative constant factor. The Fourier

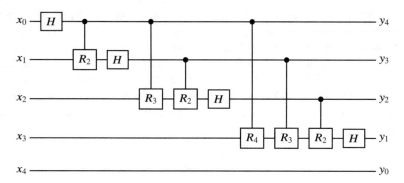

Fig. 3.2 Quantum circuit for the Fourier transform over four qubits with the qubits labeled. Notice how the order of the bits in the circuit is reversed, thus, strictly speaking to complete F, the reordering also needs to be compiled. However, we omit it because it is efficient like in classical computation, and omitting it highlights the modularity of the circuit. Finally, all control phases commute with each other and can thus be moved around, except across the Hadamard gates that do not commute with either side of the controlled phase in this way

transform is thus efficient. If it were to be simulated classically, it is known that even Fast Fourier Transform (FFT) implementations of the d dimensional Fourier transform must use $O(d)$ multiplications [Win78], which in the case of n qubits translates to $O(2^n)$ multiplications.

However, to make a fair comparison of the classical and quantum Fourier transforms, it must be noted that the QFT cannot be used to efficiently compute the classical Fourier transform of a classical vector. Just like we cannot simply measure the coefficients of the input state, we cannot measure the coefficients of the output either, and therefore not the effect of the QFT. For this reason, the implementation of the QFT is not an example of quantum speedup, as it is not solving the same computational problem as the DFT. Algorithms with a quantum speedup that make use of the QFT, as we will explain later, also make use of interference behavior in order to suppress the coefficients of unwanted outcomes and obtain a readable classical calculation.

3.4   Oracle and Promise Problems

Computational problems (i.e., computing a function) are not the only type of problems in complexity theory. Two elements that add variety are oracles and promises.

A promise is a restriction on the sets of inputs that need to be computed correctly. Outside these inputs, the behavior of the function is undefined, and an algorithm implementing it can output anything. It relaxes the requirements for the algorithms and can actually lower the complexity of a problem. Promises can also be made for the oracle problems explained below.

An oracle is a way to model finite functions as inputs to a problem, in a way that only evaluations to the functions are given and not, for example, the circuit implementing it. This is referred to as having oracle or black-box access to the function. In the circuit model, an oracle takes a given finite function f and provides a gate computing it. For a quantum oracle, the choice is not unique and a typical one is

$$CX^f := \sum_x |x\rangle\langle x| \otimes X^{f(x)} = \sum_{x,y} |x, y + f(x)\rangle\langle x, y|, \tag{3.2}$$

where x lies in the domain and y in the codomain (we assume the codomain has an addition with inverse operation). To distinguish it from the other oracles below, we sometimes call this the sum oracle.

Notice that while f might not be one-to-one, $x \rightarrow (x, f(x))$ always is. CX^f is the unitary completion of the reversible function, implemented in superposition, more precisely is a unitary completion after partially defining it with the isometry $|x\rangle \rightarrow CX^f |x\rangle |0\rangle = |x\rangle |f(x)\rangle$, which only defines CX^f on input $|0\rangle$ for the second qubit. CX^f is sometimes said to implement f *coherently*.

Another possible oracle is $|x\rangle |0\rangle = |x\rangle |y \cdot f(x)\rangle$, which we may call the multiplication oracle. Particularly important is the phase kickback, or phase oracle

$$Z_f := \sum_x \omega^{f(x)} |x\rangle\langle x|. \tag{3.3}$$

It can be obtained by acting with CX^f on an ancilla for the output in the Fourier, receiving its phase shift (hence the name), and erasing the ancilla:

$$CX^f (|x\rangle \otimes F |y\rangle) = |x\rangle \otimes X^{f(x)} F |y\rangle = |x\rangle \otimes FZ^{f(x)} |y\rangle \tag{3.4}$$

$$= \omega^{f(x) \cdot y} |x\rangle \otimes F |y\rangle = Z_f^y |x\rangle \otimes F |y\rangle, \tag{3.5}$$

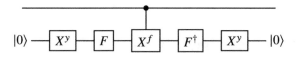

where the ancilla can be safely removed because its state does not depend on x even after the computation. Thus, the ancilla will stay in product with the input qubits even when acting on a superposition input state. More precisely,

$$CX^f (|\psi\rangle \otimes F |y\rangle) = \sum_x c_x Z_f^y |x\rangle \otimes F |y\rangle = Z_f^y |\psi\rangle \otimes F |y\rangle. \tag{3.6}$$

A single application of Z_f is obtained for $y = 1$:

$$Z_f = \left(\mathbb{1} \otimes \langle 1| \, F^\dagger\right) \cdot CX^f \cdot \left(\mathbb{1} \otimes F \, |1\rangle\right). \tag{3.7}$$

Namely, the ancilla can be produced and returned to the zero state after the query without affecting the computation.

If the output is encoded in n bits as $f(x) = f_1(x) \ldots f_n(x)$ and stored in n qubits on the quantum system, then the binary way of defining the oracle CX^f is using the binary sum

$$CX^f := \sum_x |x\rangle\langle x| \otimes \left(X^{f_1(x)} \otimes \cdots \otimes X^{f_n(x)}\right) = \sum_{x,y} |x, y \oplus f(x)\rangle\langle x, y| \, , \tag{3.8}$$

where X is now the Pauli X. While we still have $CX^f \, |x\rangle \, |0\rangle = |x\rangle \, |f(x)\rangle$ like with the general version of the oracle, the generated phase oracle is different. The appropriate one is obtained using the Hadamard rather than the Fourier transform and results in a binary phase kickback which is the ± 1 parity of the output bits, namely

$$Z_f := \sum_x (-1)^{f_1(x) \oplus \cdots \oplus f_n(x)} \, |x\rangle\langle x| = \left(\mathbb{1} \otimes (\langle 1| \, H)^{\otimes n}\right) \cdot CX^f \cdot \left(\mathbb{1} \otimes (H \, |1\rangle)^{\otimes n}\right). \tag{3.9}$$

We may call these the binary sum and binary phase oracles to distinguish them from the other ones above, with which they will coincide if the output of f is one bit.

Remark 3.2 The phase oracles are already a display of the power of quantum computation. Notice that for the classical oracle $x \to x, f(x)$, there is no way that the input register can be modified. However, because of the interplay between the conjugate bases in quantum computation, the state of the output register can determine whether the same oracles encode its output as bit flips in the "output" qubit, or as phase flips in the "input" qudit. In particular this should make clear that in terms of the effect of the quantum oracle, there is no "input" or "output" qudit. The oracle is a nontrivial unitary transformation of the joint qudit space.

Using an oracle, usually called a *query*, will cost like a single gate from the universal gate set. Counting only the queries to the oracle leads to the definition of query complexity (for families of problems), analogous to how computation complexity is defined from gate counting. Since each query is counted in the computation, a polynomial-time algorithm will also have polynomial query complexity. Notice that while it seems that the oracle changes the universal gate set, this would lead to a contradiction with the statement that the complexity of an algorithm does not depend on the universal gate set (as we could just turn the solution into an oracle). The oracle should not be thought of as part of the universal gate set because this set must be defined before any input to the problem is given, while the oracle provides the gate after the function (the input) is given and its hidden complexity can change with the size of the input.

In case the idea of an oracle seems strange because of the complexity that it hides, we present here two possible interpretations. The oracle might model the interaction with another all-powerful party that can compute the function; we only send an input and receive the function computed on it in response. Alternatively, we can think of the oracle providing us with the whole list of evaluations, which will be exponential in the number of input bits; to compute any value, we need to access the list that will count as an operation, and a polynomial algorithm will only be able to access polynomially many elements. Even if a function is efficiently computable, there are still practical examples where its circuit gives almost the same information as the oracle access. In asymmetric key encryption, a key is made public that allows us to encrypt a message, while the decryption key is kept private. This key allows us to compile and run the encryption circuit. However, if we can extract any information about the private key from the circuit, then the scheme is not secure. The computational assumptions behind the scheme, in particular, assume that with a polynomial-time algorithm no information can be extracted from the circuit that cannot be extracted from polynomially many oracle queries.

There are various reasons for considering oracle (and promise) problems. It is sometimes easier to prove speedups and separations for oracle problems; in fact, exponential speedup is proven for Simon's problem below, for example. Oracles also allow us to focus and possibly prove the efficiency of the computational part of the algorithm that does not depend on the function, which in turn allows us to gain useful insight for further applications or generalizations. Namely, they provide a template/abstraction layer for other algorithms that perform the same computation on different instances of the oracle. For example, Shor's algorithm is an application of an oracle algorithm for period finding, where the oracle is the modular exponential. The period-finding algorithm eventually was generalized to an oracle algorithm for finding hidden subgroups. Last but not least, the oracle might represent a function that is simply expensive, and we might want to minimize the amount of runs of such a function.

Because the computational complexity of the function is hidden, algorithms with different oracles are incomparable. For example, algorithms with oracle access to the discrete logarithm cannot be compared to ones with oracle access to solve the traveling salesman problem. In particular, the tricky comparison between classical and quantum algorithms might become unfair when additionally comparing classical and quantum oracles, and sometimes a different classical oracle can nullify the quantum speedup (e.g., for Deutsch–Jozsa and Simon's algorithm below), opening the argument that the speedup is simply due to the quantum oracle giving more information than the classical one. Strictly speaking, to make the quantum oracle comparable to the classical one, it should be measured after the query, and this would indeed nullify its power.

Still, there is one argument that justifies the comparison between classical and quantum oracles. As mentioned before, any efficient classical algorithm can be implemented efficiently and coherently as a quantum algorithm. In particular it means that the complexity of CX^f is not larger than the complexity of f, and thus the two oracles hide the same complexity.

Remark 3.3 Any efficiently computable classical function is also efficient as a quantum oracle.

3.5 Interference: Balanced Functions

The goal of the following algorithms is to use the balanced function problem, a simple oracle and promise problem, as a first example of quantum computation, and display the use of the Fourier transform and interference, a building block in quantum computation, in order to get a speedup.

We are given oracle access to a binary function with the promise that the function is either constant or balanced. The input is a function from n bits to a single bit $f : \{0, 1\}^n \rightarrow \{0, 1\}$, and thus the constant functions are either the 0 or the 1 function. A balanced function is a function such that the number of inputs that map to zero is the same as those that map to one, namely, $|f^{-1}(0)| = |f^{-1}(1)| = 2^{n-1}$. Such a problem has no known application, its only purpose is to display the power of quantum computation and compare it to classical computation.

Classically, the optimal strategy is to query the oracle and conclude that f is balanced or constant depending on whether all the queried values are constant. In $n = 1$, both inputs must be queried to determine if f is balanced. For larger n, the query complexity depends on whether we require an exact algorithm, or if we allow for an approximate answer from a randomized algorithm. For each possible input f, this decision problem creates a hypothesis testing scenario where the null hypothesis h is "f is constant" and, as shown in Fig. 3.3, from an exact algorithm we require the answer to be correct with certainty, while from the randomized algorithm we require the answer to be correct within some error probability ε, where the probability is computed over the randomness used by the algorithm. Clearly, an exact algorithm will have to query the function at least 2^{n-1} times; otherwise, there will always exist some balanced function that will pass as a constant function. However, if we choose the input to the queries uniformly at random, then, if the function is balanced, the probability of always hitting inputs with a constant output decreases exponentially, and more precisely, it is bounded by 2^{-k} after $k + 1$ random queries. If the function is constant, the guess will still be correct with certainty after $k + 1$ queries (Fig. 3.3).

As we can see for classical algorithms already, the query complexity may vary a lot depending on our requirements. The quantum algorithm will have better query complexity than the exact classical algorithm, but it will not improve over the approximate one that can already solve the problem in constant query complexity ($\log 1/\varepsilon$ to be precise).

Fig. 3.3 Requirement on the probabilities for an exact solution of the problem (top) and for an ε-approximate solution of the problem (center) together with the probabilities achieved by the approximate algorithm using $k+1$ queries (bottom). f is fixed, and the probabilities are computed over the randomness used by the algorithm

truth guess	f constant (h true)	f balanced (h false)
f constant (accept h)	1	0
f balanced (reject h)	0	1

truth guess	f constant (h true)	f balanced (h false)
f constant (accept h)	$> 1 - \varepsilon$	$< \varepsilon$
f balanced (reject h)	$< \varepsilon$	$> 1 - \varepsilon$

truth guess	f constant (h true)	f balanced (h false)
f constant (accept h)	1	$< \dfrac{1}{2^k}$
f balanced (reject h)	0	$> 1 - \dfrac{1}{2^k}$

3.5.1   Deutsch Algorithm

We first consider the simplest case of $n = 1$, namely of functions $f : \{0, 1\} \to \{0, 1\}$ from one bit to one bit, and the general case will then become almost straightforward. We will use the phase oracle; the usual treatment uses the sum oracle to explicitly construct the phase oracle, but using the phase oracle directly allows for a more concise treatment, and allows us to focus on interference and phase kickback, the two fundamental elements that make the algorithm work, separately. We have explained the phase kickback already, which is implicit in the phase oracle, and now the focus will be on how interference (after using the phase oracle) allows us to obtain the result with a single query.

The algorithm is extremely simple. We initialize the input state, Hadamard transform, query the oracle, Hadamard transform again, and measure in the computational basis:

$$|0\rangle\!-\!\boxed{H}\!-\!\boxed{Z_f}\!-\!\boxed{H}\!-\!\boxed{Z}\!=\ .$$

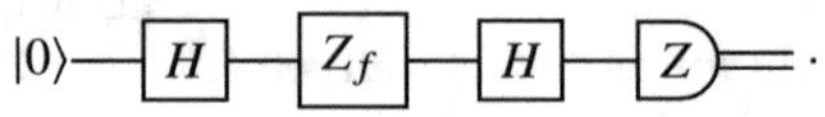

The goal of the first Hadamard is to simply generate a uniform superposition of all inputs. The oracle then changes the phases of the computational basis in the superposition:

$$Z_f H |0\rangle = \frac{1}{\sqrt{2}}\left((-1)^{f(0)}|0\rangle + (-1)^{f(1)}|1\rangle\right) \tag{3.10}$$

$$= \frac{1}{\sqrt{2}}(-1)^{f(0)}\left(|0\rangle + (-1)^{f(0)\oplus f(1)}|1\rangle\right) = (-1)^{f(0)}H|f(0)\oplus f(1)\rangle. \tag{3.11}$$

We can see that the resulting state is already encoding the answer, only in the conjugate basis, up to a global phase (that we can drop). The last Hadamard simply allows us to measure it in the computational basis:

$$HZ_f H|0\rangle = HH|f(0)\oplus f(1)\rangle = |f(0)\oplus f(1)\rangle. \tag{3.12}$$

The effect of interference becomes more explicit if we actually compute the effect of the second Hadamard on the state:

$$H\left(|0\rangle + (-1)^{f(0)\oplus f(1)}|1\rangle\right) = \frac{1}{\sqrt{2}}\left(|0\rangle + |1\rangle + (-1)^{f(0)\oplus f(1)}(|0\rangle - |1\rangle)\right), \tag{3.13}$$

where we can see that by allowing the phase to be applied in superposition without measurements, we get constructive and destructive interferences toward different basis states depending on the parity. The generalization to larger n is now straightforward.

3.5.2 Deutsch–Jozsa Algorithm

The algorithm is the same; we can still solve the problem exactly with one query, and only the analysis will be slightly different. Instead of a single bit of input, we now have n bits; therefore, we initialize and Hadamard transform n qubits:

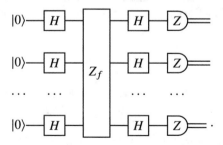

For general binary functions from any finite set with an even number of elements, we would use a qudit:

$$|0\rangle \boxed{F} \boxed{Z_f} \boxed{F^\dagger} \boxed{Z} = .$$

The state after the query is

$$|\psi_f\rangle := Z_f(H\,|0\rangle)^{\otimes n} = \frac{1}{\sqrt{2^n}} \sum_x (-1)^{f(x)} |x\rangle. \tag{3.14}$$

We do not actually need to compute the explicit state after the last Hadamard transform, and we only need to notice that if f is constant, then $|\psi_f\rangle = (H\,|0\rangle)^{\otimes n}$, otherwise it is orthogonal to $(H\,|0\rangle)^{\otimes n}$. The first claim is clear because if f is constant then the oracle applies to only a global phase; the second claim follows from

$$(\langle 0|\,H)^{\otimes n}\,|\psi_f\rangle = \frac{1}{2^n} \sum_{x,y} \langle y|\,(-1)^{f(x)}\,|x\rangle = \frac{1}{2^n} \sum_x (-1)^{f(x)}, \tag{3.15}$$

which will be zero if f is balanced because the sum will contain an equal number of $+1$s and -1s (destructive interference). The sum will be either ± 1 if f is constant (constructive interference), implying that the states are the same as argued above. Since unitary transformations in particular preserve orthogonality, the last Hadamard transform will either produce $(|0\rangle)^{\otimes n}$ if f is constant, or any other orthogonal state if f is balanced (even though not necessarily a basis state). We simply conclude that f is constant if all the measurements return 0 and balanced otherwise, giving a correct answer with probability 1 (because being orthogonal to $(|0\rangle)^{\otimes n}$ means in particular that 0^n cannot be an outcome of the basis measurement).

3.5.3   Bernstein–Vazirani Algorithm

Computing whether a function is balanced or not does not have known applications. However, balanced functions include, for example, linear binary functions, namely, functions of the form

$$f_a(x) = a \cdot x = a_1 x_1 \oplus \cdots \oplus a_n x_n. \tag{3.16}$$

For this subclass, the final measurement actually gives us a and determines the function uniquely with a single query. Indeed, we have

$$H^{\otimes n} Z_{f_a} H^{\otimes n}\,|0\rangle = H^{\otimes n} \sum (-1)^{a \cdot x}\,|x\rangle = H^{\otimes n} H^{\otimes n}\,|a\rangle = |a\rangle, \tag{3.17}$$

which is due to the oracle actually being a phase flip $Z_{f_a} = Z^a = Z^{a_1} \otimes \cdots \otimes Z^{a_n}$ that becomes X^a under the Hadamard conjugation. In contrast, classically, we need

n queries to find a, because we must query at least a basis of the bit vector space to identify any linear function uniquely.

We have seen how quantum computation allows us to solve this problem with a single query for any n, while any classical algorithm must use at least two queries. In the case of an exact algorithm, the speedup is exponential. However, while exact classical computation is possible, exact quantum computation will never be achieved in practice. This is because the space of quantum computation is continuous and noise is unavoidable, so the best we can hope for is arbitrary precision. Thus, if we assume that only approximate quantum algorithms exist, we should compare them with approximate classical algorithms only, and the quantum speedup will disappear. Still, approximating a solution does not always make a problem easier, and thus finding a problem with a quantum speedup for an exact solution is a first step toward proving a practical speedup.

Ideally we would also like to argue that the speedup is due to using quantum states and operations to process the information of the oracle query. However, it is sometimes possible to devise classical-like algorithms that provide more information than the standard classical oracle and achieve the same speedup, as is the case for the balanced functions problem [JL17]. In this case, it can be argued that the speedup is not algorithmic, meaning coming from the processing part of the algorithm, but only from the oracle itself, capable of giving more or a different type of information.

3.6 Measurements: Hidden Subgroups

So far we have seen algorithms that have avoided measurements in order to not disturb the computation. Namely, we have seen that by delaying the measurement and taking advantage of interference and the Fourier transform, we can compute some properties more efficiently. We have also seen the clean use of ancillas in the implementation of the phase oracles, where clean means that the ancilla was returned/left into a state decoupled from the computation, such that measuring or erasing it would not affect the state of the computation further.

With the next algorithm, we will see how measurements or erasures of ancillas that are still entangled with the whole state can actually be used as part of the computation. The first step is recalling that, in broad terms, a measurement destroys superposition only on the information it collects. As an example, to be more precise, consider the state (up to normalization)

$$|0\rangle |0\rangle + |1\rangle |1\rangle + |2\rangle |+\rangle + |3\rangle |-\rangle = \left(|0\rangle + \frac{1}{\sqrt{2}} |2\rangle \right) |0\rangle + \left(|1\rangle + \frac{1}{\sqrt{2}} |3\rangle \right) |1\rangle .$$
$$(3.18)$$

In the superposition, the four states of the first system are orthogonal, and thus measuring them will collapse the state of the second into $|i = 0, 1\rangle$ or $|\pm\rangle$.

However, the four states of the second system are not orthogonal, and thus measuring the computational basis $|i = 0, 1\rangle$ or the conjugate basis $|\pm\rangle$ will not collapse the first system into $|i = 0, \ldots, 3\rangle$. For example, the computational basis measurement will collapse the superposition outside the parenthesis on the right hand side, but not inside, and collapse the first system onto $|0 + i\rangle + \frac{1}{\sqrt{2}} |2 + i\rangle$ upon outcome $i = 0, 1$. This can be exploited to do co-set state preparation in hidden-subgroups and period-finding problems as we will explain now.

In period finding, we are given a finite function $f : \mathbb{Z}_d \to S$ (for some output set S that is not particularly relevant for the computation), with the promise that f is periodic and the problem is to find the period. Namely, we are promised that for some p in $\mathbb{Z}_d$ that we need to find, we have

$$f(x + hp) = f(x) \qquad \forall\, x \in \mathbb{Z}_d,\ h \in \mathbb{N} \tag{3.19}$$

and $f(x) \neq f(x')$ otherwise. This actually defines an equivalence relation among the elements of $\mathbb{Z}_d$. Because of the periodicity condition, the non-empty preimages $f^{-1}(y) = \{\, x \in \mathbb{Z}_d : f(x) = y \,\}$ have all the same size, and the preimage containing zero, namely, $H := f^{-1}(f(0)) = \{\, hp : h \in \mathbb{N} \,\}$, is actually close under addition as it is composed of all the elements of the form $h = mp$. More precisely, H is a subgroup of the group $\mathbb{Z}_d$ (with addition as group operation), and the preimages are the co-sets of H in $\mathbb{Z}_d$. There will be $d/|H| = |f(\mathbb{Z}_d)|$ (the number of different outputs) disjoint co-sets that coincide with the non-empty preimages. Each co-set can be obtained by summing a single element of the co-set, a representative, with every element in H, namely, for $y \in f(\mathbb{Z}_d)$, we have $f^{-1}(y) = \{\, x_y + hp : h \in \mathbb{N} \,\}$ and $f^{-1}(y) = x_y + H$.

The hidden-subgroup problem is the generalization of period finding. We only need the input set to have a binary operation like the sum in order for the period to be well defined, namely, for the input set to be a group G. Efficient quantum algorithms exist when G is commutative (Abelian). However the general case is still open. Given $f : G \to S$, we have the promise that there exists a subgroup H, such that f is constant in the co-sets, but distinct for different co-sets. For a periodic function f, the period is just an element p of G, in which case the subgroup H is commutative (Abelian). The subgroup H is the co-set of the identity element (0 in the case of period finding), with all the co-sets again being of the form $g + H = \{\, g + h : h \in H \,\}$ and the number of different co-sets being the number of distinct output values $|f(G)| = |G|/|H|$. The space of co-sets denoted by G/H is also a group, since $H + H = H$, and thus $(g_1 + H) + (g_2 + H) = (g_1 + g_2) + H$.[6]

---

[6]We use the addition symbol even though it is customary to write non-commutative operations as products.

3.6.1 Co-set States

The standard algorithms to solve Abelian hidden-subgroup problems consist of using interference (as shown for the balanced functions) on co-set states rather than the uniform superposition of all states. For each co-set $g + H \in G/H$, the co-set state is defined as

$$|g + H\rangle := \frac{1}{\sqrt{|H|}} \sum_{h \in H} |g + h\rangle. \tag{3.20}$$

Using the oracle for f and a measurement, we can generate one of these states uniformly at random. We simply measure the output of the oracle:

$$(\mathbb{1} \otimes \langle y|) \cdot CX^f \cdot (F |0\rangle \otimes |0\rangle) = \frac{1}{|G|} \sum_{g \in G} |g\rangle \otimes \langle y|f(g)\rangle \tag{3.21}$$

$$= \frac{|H|}{|G|} \frac{1}{|H|} \sum_{g \in f^{-1}(y)} |g\rangle \tag{3.22}$$

$$= \frac{1}{|G|/|H|} \frac{1}{|H|} \sum_{g \in g_y + H} |g\rangle \tag{3.23}$$

$$= \frac{|H|}{|G|} |g_y + H\rangle \tag{3.24}$$

as displayed in Fig. 3.4. However, the output of the measurement is not used, and the result of the computation is independent of the choice of the co-set state. The algorithm could be run on the density state

$$\frac{|H|}{|G|} \sum_{g + H \in G/H} |g + H\rangle\langle g + H|. \tag{3.25}$$

We only need to show that the outcome does not depend on g.

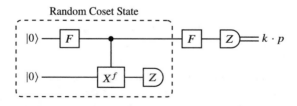

Fig. 3.4 Algorithm for period finding based on the generation of co-set states (boxed part), obtained by collapsing the state with the bottom measurement. The outcome of the measurement is not used, and thus it is equivalent to simply trace/erase the state. For the hidden-subgroup problem, the only difference is that F is the Fourier transform of the group

3.6.2   Period-Finding Algorithm

The final step is to use interference to extract the information about H from the co-set states. Since we only want to illustrate the idea of the algorithm, and the quantum part of the algorithm is the same for period finding and hidden-subgroup finding, we will limit ourselves to the more simple period-finding problem. To simplify further, we will also assume that d is a multiple of p and refer to [NC10] and [Pre98b] for the more detailed analysis and more general cases requiring additional classical post-processing. We thus have $|H| = d/p$ and $H = \{hp : h \in \mathbb{Z}_{|H|}\}$.

We focus on a single co-set state and compute the result showing how the choice of the co-set state only contributes an irrelevant phase. The phase is not global because it differs between different co-sets, but it cancels out nonetheless because the various co-set states are not in superposition, but rather in a classical mixture (due to the first measurement). More precisely, for any co-set $g + H \in \mathbb{Z}_d/H$, we have that the computational basis measurement after the Fourier transform has the probabilities

$$|\langle x| F |g + H\rangle|^2 = \left| \frac{1}{\sqrt{|H|}} \sum_{z \in \mathbb{Z}_d, h \in \mathbb{Z}_{|H|}} \frac{1}{\sqrt{d}} e^{-i \frac{2\pi}{d}(g+hp)z} \langle x|z\rangle \right|^2 \qquad (3.26)$$

$$= \frac{1}{d} \frac{1}{|H|} \left| \sum_{h \in \mathbb{Z}_{|H|}} e^{-i \frac{2\pi}{d}(g+hp)x} \right|^2 \qquad (3.27)$$

$$= \frac{1}{p} \frac{1}{|H|^2} \left| \sum_{h \in \mathbb{Z}_{|H|}} e^{-i \frac{2\pi}{|H|} hx} \right|^2 \qquad (3.28)$$

$$= \frac{1}{p} \left| \frac{1}{|H|} \sum_{h \in \mathbb{Z}_{|H|}} e^{-i \frac{2\pi}{|H|} h(x \bmod |H|)} \right|^2 \qquad (3.29)$$

$$= \frac{1}{p} \delta_{0,(x \bmod |H|)}. \qquad (3.30)$$

Therefore we get an x uniformly distributed over the multiples of $|H|$. Running the algorithm a few times will return $|H|$ as the greatest common divisor of the results, from which we can then compute p. Notice that $|H|$ is actually the inverse $1/p \bmod d$, just like the Fourier transform of real p-periodic function will be a $\frac{1}{p}$-periodic function. In the general case the algorithm will indeed return uniform multiples of $1/p$, from which p is recovered with a classical algorithm.

Classically, to find p we need to at least find a collision, namely, two queries with the same value. The number or comparison that can be done with k queries is at most k^2, which means that we need approximately $\sqrt{p}$ queries to get an efficient scaling of the success probability. If p scales as a power of d, then $\sqrt{p}$ will still scale superpolynomially in the number of bits $\log d$, and thus the problem cannot be solved efficiently.

3.6.3 Simon's Algorithm

In Simon's problem, the period, or hidden subgroup, is generated by the binary sum of n bitstrings. More precisely, we are given a function $f : \mathbb{Z}_2^n \rightarrow \mathbb{Z}_2^n$ with the promise that there exists an $a \in \mathbb{Z}_2^n$ such that $f(x) = f(x \oplus a)$ for all x and $f(x) \neq f(x')$ unless $x \oplus x' = a$, and thus all the preimages have exactly two elements. The goal is to find the unknown binary period a. In this case, because only two elements can have the same output value, we need approximately $\sqrt{2^n}$ classical queries to find a collision and determine a.

Quantum-mechanically, we will need approximately n queries. First observe that since the co-sets all have two elements, all the co-set states are of the form

$$|\bar{x}\rangle := \frac{1}{\sqrt{2}}(|x\rangle + |x \oplus a\rangle). \tag{3.31}$$

This is of course obtained using the binary sum rather than the sum oracle. Naturally, we also need to substitute the Fourier transform with $H^{\otimes n}$. The result, for which the derivation is analogous to the one for period finding, is that the measurement returns uniformly at random one of the 2^{n-1} bitstrings that satisfy $a \cdot x = 0$.

Now we just need to run the algorithm approximately n times to get with high probability n linearly independent answers $z_1, \ldots, z_n$ that we can use to determine a by solving the equations $a \cdot z_i = 0$. This is an exponential speedup over the classical algorithm that needs to find a collision pair using approximately $\sqrt{2^n}$ queries.

3.7 Phase Estimation

This time we will look at a purely quantum problem: measuring the phase of unitary matrices as an observable. Each unitary matrix U is a normal matrix, and thus unitarily diagonalizable and can be written as

$$U = \sum_k e^{i 2\pi \varphi_k} |\varphi_k\rangle\langle\varphi_k|, \tag{3.32}$$

where $|\phi_k\rangle$ is an orthonormal basis diagonalizing U (not unique if the phases are degenerate). While general unitary matrices are not Hermitian, and thus strictly speaking are not observables under the axioms of quantum mechanics, they can still be measured as a function of Hermitian observables.[7]

One option is to measure simultaneously the Hermitian and anti-Hermitian parts of U. The option considered in phase estimation is measuring the argument of U:

$$\arg U = \sum \varphi_k \, |\varphi_k\rangle\langle\varphi_k| . \tag{3.33}$$

Thus measuring $arg(U)$ will also give us a value for U. The argument is usually defined as $2\pi \varphi_k$, but here we define it as φ_k for simplicity. All the protocols seen so far can be rephrased using a phase estimation algorithm as is first done in [Kit95]. We can think of phase estimation as indirectly performing a measurement on the quantum operation, and thus it is not a surprise that it gives an alternative way to extract the information from the quantum oracles. Phase estimation also explains in full generality the mechanism of phase kickback that we have already used. We refer to [NC10] for examples of problems solved using phase estimation and more details about the efficient implementation. For simplicity, here we will do the analysis using qudits rather than qubits.

For given U, the controlled unitary CU controlled on a qudit is defined as

$$CU = \sum_{i\in\mathbb{Z}_d} |i\rangle\langle i| \otimes U^i . \tag{3.34}$$

Using the phase kickback mechanism, computing CU on the eigenstates translates into a power of the qudit phase flip. Formally, first we remark that we can take arbitrary powers of phase flips and bit flips. These are defined in the usual way of defining powers for diagonalizable matrices

$$Z^\alpha = \sum_k \omega^{k\alpha} \, |k\rangle\langle k| \qquad X^\alpha = \left(FZF^\dagger\right)^\alpha = FZ^\alpha F^\dagger . \tag{3.35}$$

Then we assume we have an oracle for producing an eigenvector $|\varphi\rangle$ of U with phase φ. Finally we see, as displayed in Fig. 3.5, that the algorithm indeed gives a bit flip power on $|0\rangle$:

$$(F \otimes \langle\varphi|) \cdot CU \cdot \left(F^\dagger \otimes |\varphi\rangle\right) = (F \otimes \langle\varphi|) \cdot \left(\sum |k\rangle\langle k| \otimes U^k\right) \cdot \left(F^\dagger \otimes |\varphi\rangle\right) \tag{3.36}$$

---

[7]This is more generally true for any normal operator K. This satisfies $K^\dagger K = KK^\dagger$, which implies that the Hermitian and anti-Hermitian parts $K \pm K^\dagger$ can be simultaneously diagonalized and thus K can be diagonalized with a complex spectrum. In other words, any normal operator can be measured by measuring simultaneously their Hermitian and anti-Hermitian parts.

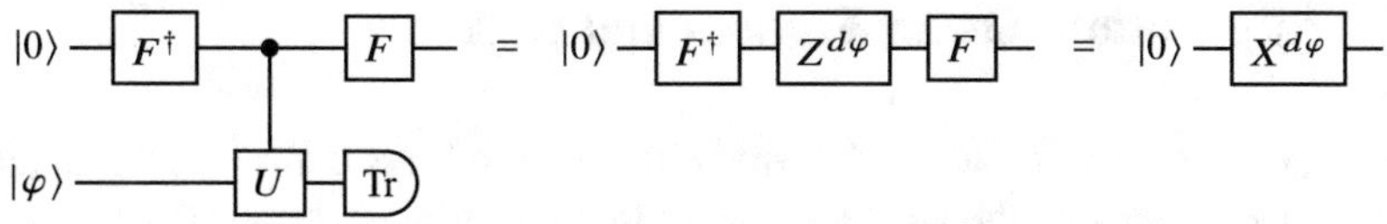

Fig. 3.5 Algorithm for phase estimation. $|\varphi\rangle$ is ideally an eigenstate of U; however, some algorithms can exploit phase estimation even with a superposition of a few eigenvectors that might be easier to produce. A clean eigenstate will not entangle with the rest of the computation, and the trace will not have any effect. If instead of $|\varphi\rangle$, a superposition $|\psi\rangle$ of eigenstates is given as input, then the algorithm will give each phase of the superposition with the corresponding probability

$$= (F \otimes \langle\varphi|) \cdot \left(\sum |k\rangle\langle k| \otimes e^{i2\pi\varphi k} |\varphi\rangle\right) \cdot \left(F^\dagger \otimes \mathbb{1}\right) \tag{3.37}$$

$$= F \cdot \sum e^{i\frac{2\pi}{d} d\varphi \cdot k} |k\rangle\langle k| \cdot F^\dagger \tag{3.38}$$

$$= F \cdot Z^{d\varphi} \cdot F^\dagger = X^{d\varphi}. \tag{3.39}$$

The main difference from what we have seen so far with other phase kickbacks is that $d\varphi$ need not be an integer. We still expect $X^{d\varphi} |0\rangle$ to be close to a computational basis state, but since non-integer values do not exist as a basis measurement outcome, we expect $X^{d\varphi} |0\rangle$ to simply be a superposition that is highly peaked around the integer values rounding $d\varphi$. Indeed the probabilities of outcome k approximate a sinc distribution around $d\varphi$:

$$p_j = \frac{1}{d} \left| \frac{\sin\left(\pi d \left(\varphi - \frac{j}{d}\right)\right)}{\sin\left(\pi \left(\varphi - \frac{j}{d}\right)\right)} \right| \sim \left| \frac{\sin\left(\pi d \left(\varphi - \frac{j}{d}\right)\right)}{\pi d \left(\varphi - \frac{j}{d}\right)} \right|^2. \tag{3.40}$$

Measuring the computational basis at this point will allow to estimate φ with $\log d$ bits of precision.

If a superposition $|\psi\rangle = \sum_k c_k |\varphi_k\rangle$ is given as input, then the global state before any measurement becomes

$$(F \otimes \mathbb{1}) \cdot CU \cdot \left(F^\dagger \otimes |\psi\rangle\right) = \sum_k c_k \cdot (X^{d\varphi_k} |0\rangle) \otimes |\varphi_k\rangle, \tag{3.41}$$

namely a superposition of the states computed previously. After the trace of the target, the state collapses to the density matrix

$$\rho = \sum_k |c_k|^2 \cdot X^{d\varphi_k} |0\rangle\langle 0| X^{-d\varphi_k}. \tag{3.42}$$

Measuring the computational basis will thus return an estimate for φ_k with probability $|c_k|^2$.

3.8   Application: Order Finding and RSA

Given a finite group M and an element m, the order of m is defined as the smallest period of m^x, namely, the smallest integer that repeating m on itself gives to the identity element 1

$$\text{Ord}_M(m) = \inf\{\, r \in \mathbb{N} \,|\, m^r = m \cdot \cdots \cdot m e \,\}. \tag{3.43}$$

We denote the group operation as a product this time and use the convention that the infimum of the empty set is $+\infty$. If the order of m is finite, then it is also a period of $x \rightarrow m^x$, and all the arguments can be taken modulo the order. Formally, $m^x = m^{x \bmod \text{Ord}_M(m)}$ for all $x \in \mathbb{N}$, and thus when doing (integer) operations on the input, it is always enough to compute everything modulo $\text{Ord}_M(m)$ because any multiple of $\text{Ord}_M(m)$ will just produce 1. When m^x is efficiently computable, we can also efficiently implement its quantum oracle, at which point the quantum algorithm for period finding with oracles becomes an efficient quantum algorithm without oracles to find the order, whenever it exists.

Whether the quantum algorithm presents an interesting speedup depends on the specific group we are considering. For example, the integers $\mathbb{Z}_d$ with modular addition also fit into the above description, but in this case exponentiation is simply integer multiplication, and finding the order is not a classically hard problem. However, m^x for the group of integers $\mathbb{Z}_d^*$ with modular multiplication is the modular exponential, and the order, denoted by $\text{Ord}_d m$, is indeed conjectured classically hard in $n = \log d$ (the number of bits needed to represent d). This is at the base of the security assumptions for today's cryptosystems and the reason why the quantum algorithm breaks the current cryptography. Both integer factoring and discrete log have efficient classical oracle solutions with a classical oracle for order finding of the modular exponential. However, both are conjectured to be classically hard, which in turn implies the conjectured hardness of order finding. The hardness assumption of integer factoring is the basis for the security of the RSA (Rivest–Shamir–Adleman) cryptosystem, while the hardness assumption of the discrete log is the basis for the security of the Diffie–Helman–ElGamal cryptosystems. The cryptosystems are constructed so that with the private key they only require computing multiplications and modular exponentials while breaking the systems, which can mean, for example, finding the key or decrypting a message, which is equivalent to the solution of some order finding problem. If the assumptions are true, the cryptosystems make it efficient to use them legitimately and inefficient to break them classically. However the quantum algorithm for order finding gives an efficient oracle to the algorithms for integer factoring and computing the discrete log, breaking the cryptosystems. We will not give a full exposition of these algorithms, and we refer to [NC10, Pre98b] for good expositions. Instead we will simply show how a single message can be decrypted from RSA without recovering the private key, since this can be done with period finding alone.

To use the modular exponential in cryptography, we need it to be invertible, namely, we need to guarantee that the discrete log exists. For this and the period to be finite, we first need the base to be invertible under multiplication, namely, it must be an element of the multiplicative group $\mathbb{Z}_d^*$, which is defined as the set of integers in $\mathbb{Z}_d$ that are coprime with d (have no common divisor). Secondly, we need the exponent to be invertible. However this will not be the same multiplicative group since $m^x \mod n = m^{x \mod n} \mod n$ is generally wrong. The integer of the exponent and the one of the base are different instances of modular arithmetic (the exponent is always an integer for any group).

By the Euler theorem, it is known that $\phi(d) = |\mathbb{Z}_d^*|$ is a period of m^x for every element of the multiplicative group m, which means that $\phi(d)$ is always a multiple of the period. The actual least common multiple of all periods is sometimes even smaller than $\phi(d)$ and is known as the Carmichael function $\lambda(d)$. Thus it always holds that $\mathrm{Ord}_d(m)|\lambda(d)|\phi(d)$, where $a|b$ means that a divides b. Both the Euler and Carmichael functions can be computed efficiently knowing the factors of n, but they are believed to be hard to compute, and it is believed to be hard to even say anything about their factors without the factors of n.

Based on all the assumptions mentioned so far, the RSA key generation, encryption, and decryption work as follows:

- To generate the private and public key:

  - the number of bits n is chosen;
  - two large primes p and q of approximately n bits size are chosen at random;
  - a number and its inverse x and $1/x$ are chosen at random in $\mathbb{Z}_{\lambda(pq)}^*$;
  - (x, pq) is published as the public key and $(1/x, pq)$ is kept as the secret key.

- To encrypt a message m, compute the ciphertext as $c = m^x \mod n$.
- To decrypt a cyphertext c, compute the message as $m = c^{1/x} \mod n$.

We now fix a message m and assume that we only know the public key and c. To break the encryption, we first note that because x was chosen in $\mathbb{Z}_{\lambda(pq)}^*$, we have $\{ c^y = m^{xy} : y \in \mathbb{Z}_{\lambda(pq)}^* \} = \{ m^y : y \in \mathbb{Z}_{\lambda(pq)}^* \}$, and thus m and c have the same periods. We thus run the period-finding algorithm on c^y, and with the returned period, we compute an inverse x' for x that will decrypt m. However we cannot directly use this inverse to decrypt other messages because it might not be an inverse in $\mathbb{Z}_{\lambda(pq)}^*$.

3.9 Grover's Search

We will now present one last quantum algorithm for solving classical problems (the next section, Quantum Simulation will address quantum problems). The feature that stands out with respect to the previous algorithms is that even though it is an oracle algorithm, Grover's search potentially speeds up almost all practically relevant problems. The drawback is that it only provides, at most, a quadratic speedup.

When we think of any search as a binary function, the input is our possible candidate and the output of the function tells us whether it is what we are searching for or not: encoded in 0 (accept) or 1 (reject). We assume that the inputs are encoded in n bits and that we are given the search function $f : \mathbb{Z}_2^n \rightarrow \mathbb{Z}_2$. The search function separates the input space of size $N = 2^n$ into the two sets of accepted and rejected elements $F_i = f^{-1}(i), i \in 0, 1$. The search problem is finding an accepted string $x \in F_1 \subseteq \mathbb{Z}_2^n$, given an oracle for f. We will also assume without loss of generality that $|F_1| < |F_0|$, which can otherwise be achieved by adding a single control bit, and that when equal to one, sends all the inputs to F_0 and computes f on the rest of the bits when equal to zero.

Most classical computational problems, even the ones that are believed to be hard to solve, have an efficient search function. Namely, a candidate solution can be checked to be a solution in polynomial time. This is, for example, the case of the discrete log and factoring. If we are given a possible solution, we can efficiently compute the modular exponential or the product to check its validity.[8] Problems where checking solutions is hard present challenges, even to find applications. Therefore for most practically relevant problems, the oracle required by Grove's search is efficiently implementable. Since a search algorithm must work for any f, for which there is no promise, there is no structure of f to be exploited, and thus a classical algorithm cannot do better than evaluating approximately $|F_1|/N$ random inputs if we require it to perform equally well on all search functions. The quantum algorithm will need approximately $\sqrt{|F_1|/N}$, which turns out to be optimal [BBBV97]. Since Grover's search can potentially give a quadratic speedup to searching cryptographic keys, but no more, doubling the key size of cryptographic schemes will maintain their computational security against brute force attacks in the presence of quantum computers [Ber10].

We define the Z operator corresponding to f as

$$Z_f = \sum (-1)^{f(x)} |x\rangle\langle x| . \tag{3.44}$$

The uniform superpositions over F_0 and F_1 are immediately checked to be eigenstates of this gate. What is more, we can write the full uniform superposition as a superposition of these two eigenstates

$$|F_k\rangle := \frac{1}{|F_k|} \sum_{x \in F_k} |x\rangle \quad \text{for } k \in \{0, 1\} \tag{3.45}$$

---

[8]The problems that can be solved efficiently form the complexity class P. Those problems for which a claimed solution can be checked efficiently form the complexity class NP, containing P and most of the practically relevant problems. One way to construct problems not in NP is to try to count the solutions to problems in NP, or even P [Val79].

$$|+^n\rangle := H^n |0^n\rangle = \sum_{k=0,1} \sqrt{\frac{|F_k|}{N}} |F_k\rangle . \tag{3.46}$$

The algorithm consists of a repetition of the phase oracle Z_f, with the next gate defined below. We define *Grover's inversion gate* as a Hadamard-rotated n-qubit controlled phase gate, only instead of changing the phase when all qubits are one, it is changed when all qubits are zero,

$$Z_{+^n} := H^n \left[-|0^n\rangle\langle 0^n| + \left(\mathbb{1} - |0^n\rangle\langle 0^n|\right) \right] H^n = \mathbb{1} - 2|+^n\rangle\langle +^n|, \tag{3.47}$$

which by construction acts as the identity outside of the subspace spanned by $\{|F_0\rangle, |F_1\rangle\}$. The algorithm will consist of repetitions of

$$G_f := -Z_{+^n} Z_f \tag{3.48}$$

on the input state $|+^n\rangle$. We omit the derivation, but it can be checked that G_f acts as a rotation in the $\{|F_0\rangle, |F_1\rangle\}$ subspace. Namely,

$$G_f = \left[\frac{1}{\sqrt{N}} \begin{pmatrix} \sqrt{|F_0|} & -\sqrt{|F_1|} \\ \sqrt{|F_1|} & \sqrt{|F_0|} \end{pmatrix} \right]^2, \tag{3.49}$$

which by defining $\cos\theta = \sqrt{\frac{|F_0|}{N}}$ or $\sin\theta = \sqrt{\frac{|F_1|}{N}}$, we can write concisely as $G_f = \left[R_f(\theta)\right]^2 = R_f(2\theta)$. In particular, it becomes evident that $|+^n\rangle = R_f(\theta)|F_0\rangle = G_f^{\frac{1}{2}}|F_0\rangle$.

At this point, we need to know the number of solutions to the search problem; namely, we need to know θ, and thus $|F_1|$. If this is not known, it can be measured using phase estimation on G_f, a procedure also known as Grover counting. Since G_f diagonalizes to $\mathrm{diag}\left(e^{i2\theta}, e^{-i2\theta}\right)$ in the $\{|F_0\rangle, |F_1\rangle\}$ subspace, any input state for phase estimation in that subspace, with $|+^n\rangle$ being an obvious efficient choice, will return $\pm\theta$, and thus $|F_1|$ (recall it is smaller than $|F_0|$ by assumption).

Knowing $|F_1|$ and thus θ, we can easily figure out how many times we need to apply G_f, so that

$$G_f^k |+^n\rangle = G_f^{k+\frac{1}{2}} |F_0\rangle = R_f((2k+1)\theta) |F_0\rangle \tag{3.50}$$

approximately rotates into $|F_1\rangle$, namely, $(2k+1)\theta \approx \pi/2$, at which point a basis measurement will return an element of F_1 with high probability. A quick calculation shows that the number of G_f gates is indeed approximately $\sqrt{\frac{N}{|F_1|}}$ (up to poly$(\log N) = $ poly(n) factors coming from the implementation of the gate).

3.10   Quantum Simulation

So far we have seen how quantum computers can help classical computation. However, the most natural use for a quantum computer, as initially pointed out by Feynman [Fey99], is to simulate quantum systems. Currently, the simulation of the energy level of molecules quickly becomes intractable with the number of atoms because of the curse of dimensionality, the exponential growth of the vector space needed for the simulation. However with a quantum computer, N energy levels can be simulated with an N dimensional quantum system, meaning $\log N$ qubits, and the quantum evolution with quantum unitaries, and thus gates. Under such conditions, the quantum computer would become a machine that behaves exactly like a molecule[9] (but unaffected by noise), is completely under control, and can be measured and reproduced arbitrarily, an experimental holy grail. As a demonstration, there has been a successful simulation of the energy levels of the hydrogen molecule [OBK$^+$16].

The initial challenge in programming a molecule is how to implement its unitary evolution. Quantum mechanics dictates that it will evolve accordingly to e^{itH}, where H is the Hamiltonian of the system, the observable of its energy. H and the unitary evolution will be some operator acting on all the atoms, thus all the energy levels, and all the qubits in the simulation. On the one hand, Hamiltonians are usually of the form $H = H_1 + \cdots + H_n$, where each term H_i is an interaction term, meaning a term involving only two atoms, and thus only a few qubits in the simulation, which can be exploited to decompose the unitary evolution into few-qubit unitary evolutions. On the other hand, these interaction Hamiltonians will not be commuting, and thus the global interaction is not the product of each interaction, because $e^{A+B} \neq e^A e^B$ for non-commuting A and B.

The solution to this problem and the foundation of quantum simulation is the Lie or Trotter product formula

$$e^{A+B} = \lim_{n \to \infty} \left(e^{A/n} e^{B/n} \right)^n. \tag{3.51}$$

Namely, we can simulate the global evolution by alternating the simulation of the interaction term in smaller terms.

The idea that a quantum problem can be more efficiently solved by a quantum computer is not exclusive to quantum simulation. Sometimes what hinders the possibility of an exponential speedup is the fact that the quantum algorithm must take classical values as inputs and encode them in the quantum computation. Part of the reason why quantum simulation is an exponentially better choice over classical simulation is that no such conversion is needed. In this case the classical algorithms already encounter a slow down, simply in converting the quantum input into

---

[9]Atoms and molecules have infinite energy levels, but only finitely many can be simulated. The simulated molecule will be a theoretical approximation with finitely many energy levels.

classical memory. An example where classical algorithms provide such a slowdown is in semi-definite programming for quantum problems.

Semi-definite programming is a subset of optimization problems where a Hermitian matrix X must be found that maximizes $\mathrm{Tr}\, XC$ for a certain Hermitian matrix C, subject to constraints on $\mathrm{Tr}\, XA_i$ for some other Hermitian matrices A_i. As a problem naturally stated as an optimization of an "expectation value" with constraints on other "expectation values," it finds vast application to quantum problems. However, in order to solve these instances with a classical computer, C and A_i must be stored classically. The known quantum algorithms that take these matrices as classical inputs can only achieve quadratic speedup [BS17]. However, by assuming that they are given as quantum input, the complexity reduces exponentially [BKL+19].

3.11 Other Applications

A survey of various quantum algorithms can be found in [BAB+18] and [Pre12]. We give a special mention to a few instances of hidden-subgroup problems that have been solved using quantum algorithms.

In [Hal07], Fourier sampling, as done in period finding, was shown to efficiently solve the principal ideal problem, and thus the Pell equation. The solution to the Pell equation gives, among other things, another solution to integer factoring, and the solution to the principal ideal problem breaks even more cryptosystems that were designed to resist the integer factoring algorithms.

In [EHKS14] it was shown that the phase estimation variant of period finding can be generalized to real vector spaces with oracles for functions that are periodic on a lattice (the subgroups of $\mathbb{R}^m$), are Lipschitz (do not vary too much), but that also do not vary too little. With sufficient precision, but at the same time sufficient coverage over the real numbers that depend on the basis of the lattice (the size of the period), the algorithms return an approximate basis for the dual lattice, and thus the lattice. This in turns solves a generalization of the principal subgroup problem.

Finding the unit group of algebraic number fields is believed to be hard to find classically and generalize all the problems of integer factoring, discrete logs, Pell equations, and principal ideal problems, and such problems were suggested as candidates to build post-quantum cryptographic systems. These groups have a periodic representation in a vector space of real numbers, integer units, and complex units, which can be embedded into a real vector space and satisfies these oracle conditions. The efficient algorithm for finding the period lattice over the real vector space, together with an efficient implementation of co-set states for the unit groups, gives an efficient quantum algorithm for finding the unit group of algebraic number fields.

Quantum algorithms have also led to exponential speedups of classical algorithms. The perfect example is in algorithms for recommendation systems. These algorithms, extensively used by service providers of large amounts of contents with large user bases (Youtube, Netflix, Amazon,...), are supposed to recommend an entry (a sample of a vector) knowing only a few preferences of a user (part of the preference vector), with the assumption that the incomplete collected data of users and preferences (a sparse matrix) depends only on a relatively small amount of the general user profile (the rank of the data matrix). In [KP17] a quantum algorithm was found that presented an exponential speedup over the known classical algorithms, simply by exploiting the fact that a measurement in quantum mechanics provides a sample of a vector, even if the vector (the quantum state) is unaccessible. In contrast, previous classical algorithms estimated the whole preference vector in order to provide a sample. The algorithm inspired a classical exponential speedup, and the classical algorithm of [Tan19] only presents a polynomial slowdown on the quantum algorithm. Indeed, the classical algorithm of [Tan19] avoids the computation of the preference vector and still provides a sample only using classical techniques.

3.12   Immediate Future

Beyond finding useful quantum algorithms, there is also interest in finding speedups for problems with possibly no application, but that can at least be implemented with a small number of qubits and can be used to demonstrate the quality of current devices. Current physical qubit implementations are too noisy to be useful for quantum computation at the levels of Grover's search or Shor's algorithm. Quantum error correction, a topic covered in the next section, is needed to produce a few good quality logical (virtual) qubits out of many noisy physical ones. The expected ratio of logical and physical qubits is one to 50–100 [Pre18]. Even then, the physical qubits themselves must operate above a certain noisy threshold for error correction to work [ABO08, FSG09]. Since so many physical qubits and such little noise is needed to run even the simplest algorithm on a few logical qubits, the question arises of how to assess that the intermediate goals toward building the smallest quantum computer have been achieved. In particular, the term "achieving quantum supremacy" has been coined to mark the point where even without a full quantum computer, the available physical qubits allow us to solve a problem or perform any computation much more efficiently than current supercomputers [Pre12]. Since the intermediate goal is benchmarking the quality of the qubits, the problem does not need to be useful.

Nonetheless, it has been argued that with the current intermediate technology (above 50 and up to a few hundred physical qubits) there can still be practical applications [Pre18]. Aside from providing hands-on experience that can complement theory and thus accelerate the progress of research, such intermediate quantum processors could potentially already solve certain optimization problems more efficiently (either in terms of cost or time). In particular, either noisy gate based processors or more stable quantum analog processors (that for the time being can yield comparable performance) can be used to augment the power of classical optimization algorithms. The optimization would not be done via quantum, but via classical computation, the classical optimizer would then get its response from the preparation and measurement of a quantum state and update the circuit accordingly. These optimization problems are allegedly more resilient to noise than regular quantum algorithms, and thus they might be implemented without the need of quantum error correction or particularly good gates. The trade-off, however, is that the simpler the quantum-assisted optimization algorithm is, the easier its classical simulation is, but within that trade-off might lie a quantum advantage that can be achieved with intermediate quantum computers.

Chapter 4
Quantum Information Theory

Despite many counter-intuitive mysteries and hence a feeling of fear that might strike the engineering or mathematics student upon hearing its title, the field of quantum information theory can be presented as a natural statistical theory, accommodating a more general concept of state, observation and correlation than those previously existing in information theory and statistics.

The necessity to conjure up a more general statistical theory started in the 1920s, where physicists were trying to explain phenomena such as Bose-Einstein condensation and stability of atoms with even numbers of electrons. Soon they realized that the existing statistics, such as Maxwell–Boltzmann statistics, could not assign two different points to *indistinguishable* particles, making it, in turn, impossible for two indistinguishable particles to have any correlation. In the classical world however, one does not care about this, as when necessary, all classical objects can be distinguished. Two objects might share their physical properties (the case with identical particles) such as mass, charge or spin, but they could always be distinguished given their trajectories. This possibility is ruled out by the wave-like behavior of microscopic particles such as electrons. Heisenberg's uncertainty principle famously prohibits exact measurements of an electron's trajectory and hence distinguishes it from another electron when in close enough proximity.

Quantum statistics, by moving from the classical picture in which probability distributions and random variables describe the system to a picture where that is done by vector spaces and density matrices, allow us to have two points in the set of states assigned to two indistinguishable particles. Loosely put, two density matrices can have the same set of eigenvalues, yet infinitely many different sets of eigenbases. This new statistics, as is readily obvious, calls for new information theoretic analysis by offering new possibilities. The first information-theoretic task that comes to mind is state discrimination, or hypothesis testing.

In this spirit, we present the most prominent tasks and protocols performed in the context of quantum information theory, invoking finite dimensional alphabets,

R. Bassoli et al., *Quantum Communication Networks*, Foundations in Signal
Processing, Communications and Networking 23,
https://doi.org/10.1007/978-3-030-62938-0_4

vector spaces and their properties. The purpose of this chapter is therefore, above all, to consider the limits to which the implications of quantum statistics can change information processing. The theoretical nature of this task calls for a rather formal and mathematical study, through which implications of quantum statistics are demystified, set against real world models of information processing and examined for their potential improvements to the classical picture. The result of this endeavor will be capacity theorems and existence results rather than the algorithms that achieve them. This methodology is analogous to what is done in a standard (classical) information theory course [CK81]. This is done here in two parts.

In the first part, we introduce the quantum equivalents of some basic Shannon information theoretic tasks that prove fundamental to more advanced research in the field. We also introduce tasks for which there are no classical counterparts. We will see that these tasks involve a new kind of correlation that does not exist in classical statistics, namely entanglement. We give some overview of the topics covered in the first step mentioned above.

Dense coding and teleportation. Here we present two communication protocols made possible by the use of perfect shared entanglement and channel between communication parties. Teleportation and dense coding are therefore first examples of protocols that do not have classical counterparts.

Quantum hypothesis testing. As a first asymptotic quantum statistical problem, we discuss the quantum hypotheses testing problem. We present a quantum version of Stein's lemma, which assigns an information theoretic meaning to quantum relative entropy.

Quantum source compression. As the second asymptotic information theoretic task, we introduce source compression of discrete memoryless quantum sources. We get to know quantum fidelity and entanglement fidelity as meaningful figures of merit. The optimal compression rate is determined in terms of the von Neumann entropy.

Transmission over classical-quantum and quantum channels. Here we analyze the transmission of classical messages over discrete and memoryless classical-quantum and quantum channels. We present the Holevo–Schumacher–Westmoreland theorem to determine the classical message transmission capacity. The coding theorem as well as the converse theorem can be derived as implications of quantum Stein's lemma.

Entanglement assisted classical message transmission. Here, the first example of an integrated task that is possible only in the realm of quantum theory is presented. While there is only a multi-letter capacity characterization available for message transmission over quantum channels, it turns out that when the communicating parties have shared entanglement at their disposal, we obtain a single letter characterization.

Security. We already described the advantages of physical layer service integration in the general introduction. When implementing the concept, the information-theoretical analysis is particularly important. The concept of physical layer security is becoming more attractive since it solely uses the physical properties of the channel to establish security. So regardless of what transformation is applied to the signals that are received by non-legitimate receivers, the original message cannot be reproduced. Therefore, such approaches provide so-called unconditional security and, not surprisingly, are identified by operators and national agencies as promising and important tasks for next-generation mobile networks. The analysis of information theoretic security for different models of channel uncertainty is an important research field and thus, indispensable for bringing this concept into practice. Here, we introduce the *wiretap* channel, a model in which information security is considered. We then consider tasks of secure message transmission, using this very model. The sender and legal receiver are interested in coding strategies that keep the wiretapper ignorant of the details of their communication.

Identification. We consider the task of message identification over cq channels and cqq wiretap channels, on which information theoretic security is modeled. The task of identification was first introduced by Ahlswede and Dueck, realizing that while Shannon's theory of message transmission presumes that the receiver wants to know everything about the message, in reality he may be interested only in certain aspects of it. In other words, the receiver may want to compute a function of the message. The most extreme case is that of identification: for sent message m and an arbitrary message m_0, the receiver would like to be able to answer the question *Is $m = m_0$?* as accurately as possible.

In the second part of this chapter, we take a step closer to real-world application of the tasks and topics introduced in the second section by relaxing some assumptions that no longer hold as they do in the ideal picture. Take, as an example, a communication scenario where a sender and a receiver connected via a noisy channel wish to communicate. In the ideal picture, the exact state of the channel is known to both parties, an assumption that is rarely the case in the real world. Departing from this picture, we would have to consider more general channel models that correspond to the classically known compound or arbitrarily varying channel uncertainties. We will see that in the context of quantum information theory, these models will also have to be generalized. For instance, in the context of the arbitrarily varying channel model, a jammer who wishes to destruct the communication channel will have more powers by having entangled states at his or her disposal. This kind of attack is known as the 'general attack', as it generalizes jamming attacks that are possible in the classical regime, known collectively as 'collective attacks'. Consideration of these tasks will require no more mathematical knowledge than that required for the first part of the chapter, as the framework continues to stay finite dimensional and discrete. The coding strategies that deal with such channel uncertainties will, of course, be significantly more sophisticated. It is not clear whether any of the coding strategies that are not specifically designed for these channel models are sufficient. As an example, one may consider coding strategies that are designed for the so-called one-shot regime, where the users of the channel do not have access to asymptotically many uses of the channel, and analyze their optimality for the compound and arbitrarily varying channel models.

4.1 Dense Coding and Teleportation

In this section we introduce two protocols made possible by pre-shared entanglement between communicating parties. These protocols were first proposed in 1993 by Bennett and Brassard, in collaboration with others [BBC+93].

First we start with the dense coding protocol, where a sender and a receiver communicate a message, using pre-shared entanglement. Notice that in the proceeding sections, we will introduce the task of classical message transmission over discrete memoryless quantum channels. Here however, the communicating parties have access to the rather expensive resource of entanglement, in addition to the channel. Before talking about dense coding, we give a formalized piece of evidence for the (at least within the quantum information theory community) famous claim that *shared entanglement alone does not suffice for message transmission*.

Let A denote the sending party, while B is the receiver. Assume that they share a state $\rho \in \mathcal{S}(\mathcal{K}_A \otimes \mathcal{K}_B)$. The most general way to set up a message transmission scheme for a number of M messages is to assign a c.p.t.p. map $\mathcal{E}_m : \mathcal{L}(\mathcal{K}_A) \to \mathcal{L}(\mathcal{H}_A)$ with any Hilbert space $\mathcal{H}_A$, and a matrix D_m, $0 \leq D_m \leq \mathbb{1}_{\mathcal{K}_B}$ for each $m \in [M]$, such that $\sum_{m=1}^{M} D_m = \mathbb{1}_{\mathcal{K}_B}$. Assume, that

$$\mathcal{E}_m(a) = \sum_{k=1}^{K} A_k a A_k^* \qquad (a \in \mathcal{L}(\mathcal{K}_A)) \qquad (4.1)$$

is any Kraus decomposition for $\mathcal{E}_m$. The probability that m' is received while m was sent is given by

$$p(m'|m) = \mathrm{Tr}\left[(\mathbb{1}_{\mathcal{H}_A} \otimes D_{m'})(\mathcal{E}_m \otimes \mathrm{id}_{\mathcal{K}_B}(\rho))\right] \qquad (4.2)$$

$$= \sum_{k=1}^{K} \mathrm{Tr}\left[(\mathbb{1}_{\mathcal{H}_A} \otimes D_{m'})(A_k \otimes \mathbb{1}_{\mathcal{K}_B})(\rho)(A_k \otimes \mathbb{1}_{\mathcal{K}_B})^*\right] \qquad (4.3)$$

$$= \mathrm{Tr}\left[D_{m'} \rho_B\right]. \qquad (4.4)$$

Inspection of the above chain of equalities shows that the probability of receiving m' is *independent of the sent* m. In consequence, if $m_1, m_2 \in [M]$ are any two distinct messages, and $p(m_1|m_1) \geq 1 - \lambda$ for some $\lambda \in [0, 1]$, then

$$p(m_2|m_2) = 1 - \sum_{m \neq m_2} p(m|m_2) \leq 1 - p(m_1|m_2) = 1 - p(m_1|m_1) < \lambda.$$

$$(4.5)$$

It is therefore not possible to transmit one of two messages with an error less than $\frac{1}{2}$, which can be also achieved if the receiver randomly guesses the message.

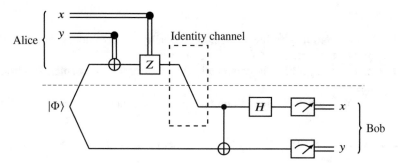

Fig. 4.1 Dense coding

Dense coding is the protocol that results from using the ideal quantum channel $\mathrm{id}_{\mathcal{H}}$ together with a pure maximally entangled state on $\mathcal{H} \otimes \mathcal{H}$. In this case, d^2 messages can be transmitted with perfect reliability, i.e. the message transmission capacity without entanglement assistance is exceeded by a factor of 2!

Theorem 4.1 (Dense Coding) *Let* $\mathcal{H}_A = \mathcal{H}_B = \mathcal{K}_B = \mathbb{C}^d$ *and* id *be the ideal channel mapping* $\mathcal{H}_A$ *to* $\mathcal{H}_B$. *There exists a family* $\{\mathcal{E}_m\}_{m=1}^{d^2} \subset \mathcal{C}(\mathcal{H}_A, \mathcal{H}_A)$ *and a POVM* $\{D_m\}_{m=1}^{d^2}$, *such that with* $\psi := \sqrt{d}^{-1} \sum_{k=1}^{d} e_k \otimes e_k \in \mathcal{H}_A \otimes \mathcal{K}_B$ *(or equivalently* $\Psi := |\psi\rangle \langle\psi| \in \mathcal{S}(\mathcal{H}_A \otimes \mathcal{K}_B)$*), for each* $m, m' \in [d^2]$,

$$p(m'|m) := \mathrm{Tr}\left[\mathrm{id} \circ \hat{\mathcal{E}}_m \otimes \hat{D}_{m'}(\Psi)\right] = \delta_{mm'} := \begin{cases} 1 & \textit{if } m = m' \\ 0 & \textit{if } m \neq m' \end{cases}. \qquad (4.6)$$

We observe that with a channel connecting the sender and receiver (see Fig. 4.1), they can reliably transmit messages. In Sect. 4.5, we consider coding strategies to transmit messages over general quantum channels using pre-shared entanglement. It can be shown that entanglement can be used as a resource to keep the communicated messages secure from a third illegal party. We will introduce the wiretap model in which information security is considered in Sect. 4.6.

We end this section by presenting another protocol, namely teleportation, that is also made possible by pre-shared entanglement. Before stating the relevant theorem, we introduce notation that goes beyond quantum channels.

Definition 4.1 (Operation) Let $\mathcal{H}, \mathcal{K}$ be Hilbert spaces. Recall that a completely positive map $\mathcal{T} : \mathcal{L}(\mathcal{H}) \to \mathcal{L}(\mathcal{K})$, which in addition is trace non-increasing, i.e.

$$\mathrm{Tr}\, \mathcal{T}(A) \leq \mathrm{Tr}\, A \qquad\qquad (A \in \mathcal{L}(\mathcal{H}), A \geq 0), \qquad (4.7)$$

is called a *(quantum) operation*. We define the short notation

$$\mathcal{C}^{\downarrow}(\mathcal{H}, \mathcal{K}) := \{\mathcal{T} : \mathcal{L}(\mathcal{H}) \to \mathcal{L}(\mathcal{K}) : \mathcal{T} \text{ is c.p., and } \forall A \geq 0 : \mathrm{Tr}\, \mathcal{T}(A) \leq \mathrm{Tr}\, A\}.$$

$$(4.8)$$

Remark 4.1

(i) By definition, a quantum channel (completely positive and trace preserving map) is an operation, i.e. $\mathcal{C}(\mathcal{H}, \mathcal{K}) \subset \mathcal{C}^{\downarrow}(\mathcal{H}, \mathcal{K})$.
(ii) Because an operation is completely positive (c.p) in particular, it admits a Kraus representation.
(iii) An operation can be always completed to be a quantum channel by adding a suitable c.p map.

Definition 4.2 (Instrument) Let $|\mathcal{X}| < \infty$. A *(quantum) instrument* is a family $\{\mathcal{T}_x\}_{x \in \mathcal{X}}$ such that

1. $\mathcal{T}_x \in \mathcal{C}^{\downarrow}(\mathcal{H}, \mathcal{K})$ for each $x \in \mathcal{X}$, and
2. $\sum_{x \in \mathcal{X}} \mathcal{T}_x \in \mathcal{C}(\mathcal{H}, \mathcal{K})$.

Definition 4.3 (One-Way LOCC Channels) Let $\mathcal{H}_A, \mathcal{H}_B, \mathcal{K}_A, \mathcal{K}_B$ be Hilbert spaces of systems under the control of communication parties A and B. A quantum channel $\mathcal{N} \in \mathcal{C}(\mathcal{H}_A \otimes \mathcal{H}_B, \mathcal{K}_A \otimes \mathcal{K}_B)$ is a LOCC channel with local operations regarding A and B and (noiseless) classical communication from A to B ($A \to B$-one-way LOCC channel), if it can be written in the form

$$\mathcal{N}(a) := \sum_{k=1}^{N} \mathcal{A}_k \otimes \mathcal{B}_k(a) \tag{4.9}$$

where $\{\mathcal{A}_k : k \in [N]\} \subset \mathcal{C}^{\downarrow}(\mathcal{H}_A, \mathcal{K}_A)$ is a quantum instrument, and $\mathcal{B}_k \in \mathcal{C}(\mathcal{H}_B, \mathcal{K}_B)$ is a quantum channel for each $k \in [N]$, where $[N] = \{1, \ldots, N\}$.

It is known that one cannot generate any state beyond the class of separable states by LOCC preparations alone. A very interesting class of protocols arises if two parties can use pre-shared entangled states in addition to local operations and classical communication. One prominent example of this class is the so-called *quantum teleportation* protocol (Fig. 4.2).

Theorem 4.2 (Quantum Teleportation) *Let $\mathcal{H}_A \simeq \mathcal{K}_A \simeq \mathcal{K}_B$ be Hilbert spaces. There exists an $A \to B$ one-way LOCC channel $\mathcal{T} \in \mathcal{C}(\mathcal{H}_A \otimes \mathcal{K}_A \otimes \mathcal{K}_B, \mathcal{K}_B)$ such that with a pure maximally entangled state $\Phi \in \mathcal{S}(\mathcal{K}_A \otimes \mathcal{K}_B)$*

$$\mathcal{T}(a \otimes \Phi) = a \tag{4.10}$$

for each $a \in \mathcal{L}(\mathcal{H}_A)$.

The quantum state a is therefore communicated perfectly with the receiver in charge of Hilbert space $\mathcal{K}_B$, using the given resources.

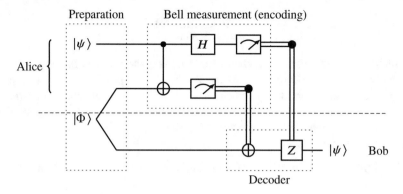

Fig. 4.2 Teleportation protocol. The dashed line separates the distant parties (Alice and Bob) and marks the need for classical or quantum communication

4.2   Quantum Hypotheses Testing: Quantum Stein's Lemma

In this section we consider the *asymmetric quantum hypothesis testing* problem. Assume an experimenter is confronted with a source which emits pairwise independent and equally prepared quantum systems. Given two a priori density matrices σ_0 (called *null hypothesis*) and σ_1 (called *alternative hypothesis*), the goal is to decide by measurements on the outputs, which preparation is present. Quantum Stein's lemma quantifies the behavior of the error of optimal tests for this task in a situation where large numbers of outputs of the systems are available for performing tests.

This task is central to quantum information theory for two reasons. Firstly, hypothesis tests also make up for good message transmission codes. This fact is already known from classical Shannon information theory. However, this relation seems to be even more important for quantum systems, as we will see in subsequent sections. Secondly, Stein's lemma allows a very simple and illuminating proof of the *monotonicity of the quantum relative entropy under completely positive and trace preserving maps* (c.p.t.p), which is notoriously hard to prove otherwise. After all, this will be our entrance to several highly nontrivial quantum entropic inequalities which are essential for proving major results in quantum Shannon theory.

The mentioned strategy to prove entropy inequalities starting from Quantum Stein's lemma is strongly inspired by the paper [BSS12], where the relatively elementary proof of the result given below can be found. The interested reader should consult that work. To formally settle the situation described above, assume we are confronted with a preparation device which emits quantum systems pairwise uncorrelated, and additionally are all being prepared according to the same density matrix. If this density matrix is γ, the mentioned properties of the preparation device ensure us that the statistical behavior of the joint quantum state of n systems (*blocklength n*) prepared is described by the density matrix

$$\gamma^{\otimes n} := \underbrace{\gamma \otimes \cdots \otimes \gamma}_{n \text{ times}}. \tag{4.11}$$

To settle the (asymmetric) hypothesis test problem, assume now that the state γ is unknown to the receiver of the systems. The receiver is provided with two a priori hypotheses in the form of density matrices ρ (*null hypothesis*) and σ (*alternative hypothesis*), and tries by measurement on the outputs to decide which of these hypotheses to accept. For a given test $\{E_0, E_1\}$ on the n-fold output system, two kinds of errors can happen

1. **First kind error:** The actual density matrix is ρ, but σ is detected. This happens with probability

$$\text{Tr}\left[E_1 \rho^{\otimes n}\right] = \text{Tr}\left[(\mathbb{1} - E_0)\rho^{\otimes n}\right]. \tag{4.12}$$

2. **Second kind error:** The density matrix is σ, but ρ is detected. This happens with probability

$$\text{Tr}\left(E_0 \sigma^{\otimes n}\right). \tag{4.13}$$

A common goal now is to determine the optimal asymptotic behavior of the second kind error for tests whose first kind error is below a threshold $\epsilon \in (0, 1)$. For this reason, we define for each $\epsilon \in [0, 1]$,

$$\beta_{\epsilon, n}(\rho, \sigma) := \inf\left\{\text{Tr}\left(a\sigma^{\otimes n}\right) : 0 \le a \le \mathbb{1}_{\mathcal{H}}^{\otimes n} \text{ and } \text{Tr}\left(a\rho^{\otimes n}\right) \ge 1 - \epsilon\right\}. \tag{4.14}$$

To formulate the quantum version of Stein's lemma, we need the following definition.

Definition 4.4 (Quantum Relative Entropy) The *quantum relative entropy* of a pair $(\rho, \sigma) \in S(\mathcal{K}) \times S(\mathcal{K})$ is defined

$$D(\rho||\sigma) := \begin{cases} \text{Tr}\left[\rho(\log \rho - \log \sigma)\right] & \text{if } \ker\sigma \subset \ker\rho \\ +\infty & \text{otherwise.} \end{cases} \tag{4.15}$$

The following theorem is the quantum theoretic generalization to Stein's lemma.

Theorem 4.3 (Quantum Stein's Lemma) *Let $\rho, \sigma \in S(\mathcal{H})$ be density matrices with $\ker\sigma \subset \ker\rho$. For each $\epsilon \in (0, 1)$, it holds*

$$\lim_{n \to \infty} \frac{1}{n} \log \beta_{\epsilon, n}(\rho, \sigma) = -D(\rho||\sigma). \tag{4.16}$$

This assertion in quantum information theory textbooks is proved in two steps.

$$\limsup_{n\to\infty} \frac{1}{n} \log \beta_{\epsilon,n}(\rho,\sigma) \leq -D(\rho||\sigma), \tag{4.17}$$

which implies together with

$$\liminf_{n\to\infty} \frac{1}{n} \log \beta_{\epsilon,n}(\rho,\sigma) \geq -D(\rho||\sigma), \tag{4.18}$$

the assertion of the theorem.

4.3   Source Compression for Memoryless Quantum Sources

We consider a *discrete memoryless quantum source* (DMQS) with generic density matrix $\rho \in \mathcal{S}(\mathcal{H})$. In this model, the state of n source outputs is the n-fold tensorial extension of the generic density matrix ρ, i.e.

$$\rho^{\otimes n} = \underbrace{\rho \otimes \cdots \otimes \rho}_{n \text{ times}}. \tag{4.19}$$

In this section we consider the extent to which the outputs of such a source can be compressed. We regard, in this sense, the system's degrees of freedom, i.e. the dimensionality of the relevant Hilbert space as a costly resource and look for source compression schemes, which map the statistics of outputs of a source to a system with a smaller number of dimensions, such that it can be recovered with negligible error.

When one aims to *perfectly* store n outputs of the DMQS ρ, one can show that a space of $(\text{rank}\rho)^n$ dimensions is necessary. This amount can in general be substantially decreased, if we allow small imprecision (error) in recovery. Figure 4.3 depicts a general source compression code for block-length n.

In order to show a coding theorem and a converse to quantify the asymptotics of the optimal compression rates, we need a sufficient performance measure to quantify the quality of recovery. We will use the *entanglement fidelity* for this purpose.

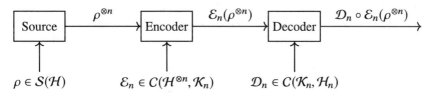

Fig. 4.3 A general source compression code for block-length n

We have introduced this measure in Sect. 2.4.2. In the following, we state the source compression theorem for DMQS. The optimal compression rate will be determined by the von Neumann entropy of the generic density matrix. We discuss some properties of this function in Sect. 4.3.

We will now determine the optimal rate for compression of a DMQS.

Definition 4.5 An (n, k)-*code for source compression* of the DMQS $\rho \in \mathcal{S}(\mathcal{H})$ is a pair $(\mathcal{E}, \mathcal{D})$, where $\mathcal{E} \in \mathcal{C}(\mathcal{H}^{\otimes n}, \mathcal{K})$, $\mathcal{D} \in \mathcal{C}(\mathcal{K}, \mathcal{H}^{\otimes n})$ are c.p.t.p. maps, and $k = \dim \mathcal{K}$. We define for each $n \in \mathbb{N}, \epsilon \geq 0$

$$K(\rho, n, \epsilon) := \min \left\{ k : \exists (n, k)\text{-code } (\mathcal{E}, \mathcal{D}) \text{ with } F_e(\rho^{\otimes n}, \mathcal{D} \circ \mathcal{E}) \geq 1 - \epsilon \right\}. \tag{4.20}$$

The following assertion is known as the source compression theorem for discrete memoryless quantum sources.

Theorem 4.4 (DMQS Source Compression Theorem) *Let* $\rho \in \mathcal{S}(\mathcal{H})$. *It holds for all* $\epsilon \in (0, 1)$

$$\lim_{n \to \infty} \frac{1}{n} \log K(\rho, n, \epsilon) = S(\rho). \tag{4.21}$$

The above theorem provides the von Neumann entropy S with an operational meaning. $S(\rho)$ is the minimal compression rate for asymptotically perfect compression of the DMQS ρ.

Some Properties of the von Neumann Entropy

We have seen that the von Neumann entropy $S(\rho)$ of a density matrix ρ has an interpretation as the optimal asymptotic rate for compression of the memoryless quantum source generated by ρ. Here we state some properties of this function.

Proposition 4.1 (Concavity of the von Neumann Entropy) *Let* $\rho_1, \rho_2 \in \mathcal{S}(\mathcal{H})$, *and* $\lambda \in (0, 1)$. *It holds*

$$S(\lambda \rho_1 + (1 - \lambda)\rho_2) \geq \lambda S(\rho_1) + (1 - \lambda)S(\rho_2). \tag{4.22}$$

Proposition 4.2 (Monotonicity of S Under Pinching Channels)
Let $P_1, \ldots, P_K \subset \mathcal{L}(\mathcal{K})$ *be mutually orthogonal projections (i.e.* $P_k^* = P_k$, $P_k P_k = P_k$, *and* $P_k P_{k'} = 0, k' \neq k$. *It holds*

$$S\left(\sum_{k=1}^{K} P_k \rho P_k \right) \geq S(\rho) \tag{4.23}$$

for all $\rho \in \mathcal{S}(\mathcal{K})$.

Proposition 4.3 (Almost-Convexity of the von Neumann Entropy) *Let* $\{\rho_x\}_{x \in \mathcal{X}} \subset \mathcal{S}(\mathcal{K})$ *be a finite family of density matrices, and* $q \in \mathcal{P}(\mathcal{X})$ *a probability distribution. Define*

$$\overline{\rho} := \sum_{x \in \mathcal{X}} q(x)\rho_x. \tag{4.24}$$

It holds

1. $S(\overline{\rho}) \leq \sum_{x \in \mathcal{X}} q(x)S(\rho_x) + H(q)$.
2. Equality in Item 1 holds if and only if supp$\rho_x \perp$ supp$\rho_{x'}$ *for all* $x \neq x'$.

Lemma 4.1 (Invariance Under Isometric Transformation) *Let* $v : \mathcal{H} \rightarrow \mathcal{K}$ *be an isometric linear map (i.e.* $vv^* = \mathbb{1}_{\mathcal{H}}$*, and* v^*v *is an orthogonal projection in* $\mathcal{K}$*). Then*

$$S(v\rho v^*) = S(\rho) \tag{4.25}$$

does hold for all $\rho \in \mathcal{S}(\mathcal{H})$. *In particular S is invariant under unitaries.*

Lemma 4.2 (Lower and Upper Bounds) *Let* $\rho \in \mathcal{S}(\mathcal{H})$. *It holds*

$$0 \leq S(\rho) \leq \log \dim \mathcal{H}. \tag{4.26}$$

Lemma 4.3 (Sub-additivity) *Let* $\rho \in \mathcal{S}(\mathcal{H} \otimes \mathcal{K})$, $\sigma_1 := \mathrm{Tr}_{\mathcal{K}}(\rho)$, $\sigma_2 := \mathrm{Tr}_{\mathcal{H}}(\rho)$. *It holds*

$$S(\rho) \leq S(\sigma_1) + S(\sigma_2), \tag{4.27}$$

where equality holds if $\rho = \sigma_1 \otimes \sigma_2$.

4.4 Message Transmission over Quantum Channels

In this section we devote ourselves to a discussion of classical message transmission over quantum channels. In Sect. 4.4.1 we discuss channel coding over a semi-classical model—channels with classical input and quantum output. We determine the message transmission of discrete memoryless channels of this type. In Sect. 4.4.2 we generalize the model to a channel with quantum input and quantum output.

4.4.1   The Discrete Memoryless Classical-Quantum Channel

In this section, we assume that the sender and receiver are connected by a transmission line, where the input is a classical symbol and the output is a quantum system. This scenario is modeled by a so-called *classical-quantum channel* (or *cq channel*), which is a map

$$V : \mathcal{Y} \to \mathcal{S}(\mathcal{K}), \tag{4.28}$$

$$y \mapsto V(y) \in \mathcal{S}(\mathcal{K}) \tag{4.29}$$

for some alphabet $\mathcal{Y}$ and Hilbert space $\mathcal{K}$. If many uses of such a channel are available in a way that the transmissions are all mutually independent, we model the transmission by the following memoryless channel model.

Definition 4.6 (Discrete Memoryless Classical-Quantum Channel) The *discrete memoryless classical-quantum channel (DMCQC)* generated by a cq channel $V :$ $\mathcal{X} \to \mathcal{S}(\mathcal{H})$ is given by the family $\{V^{\otimes n}\}_{n \in \mathbb{N}}$ where for each $n \in \mathbb{N}$ the cq channel $V^{\otimes n}$

$$V^{\otimes n}(x^n) := \bigotimes_{i=1}^{n} V(x_i) = V(x_1) \otimes \cdots \otimes V(x_n) \tag{4.30}$$

for each $x^n = (x_1, \ldots, x_n) \in \mathcal{X}^n$.

Having defined a channel model, a standard task in information theory is to give a general capacity formula which quantifies the message transmission abilities. We aim to determine the message transmission capacity of the DMCQ channels defined above.

A channel code for n uses of a classical-quantum channel is usually given by a codeword u_m for each message m. Since the outputs of the channel are quantum mechanical systems, the decoding is performed by a quantum measurement (i.e. POVM) on the Hilbert space belonging to n outputs of the channel (a general coding scheme is depicted in Fig. 4.4).

We give rigorous definitions for the coding scenario.

Definition 4.7 An (n, M) *code for classical message transmission* over the DMCQC generated by $V : \mathcal{X} \to \mathcal{S}(\mathcal{H})$ is a family $\mathcal{C} := (u_m, D_m)_{m=1}^{M}$, where $u_1, \ldots, u_m$ are words in $\mathcal{X}^n$, and $\{D_m\}_{m=1}^{M}$ is a POVM in $\mathcal{L}(\mathcal{H}^{\otimes n})$. With the shortcut $D_m^c := \mathbb{1}_{\mathcal{H}}^{\otimes n} - D_m$, we define the functions

$$\overline{e}(\mathcal{C}, V^{\otimes n}) := \frac{1}{M} \sum_{m=1}^{M} \mathrm{Tr}\left[D_m^c V^{\otimes n}(u_m)\right] \qquad \text{(average transmission error)},$$

$$\tag{4.31}$$

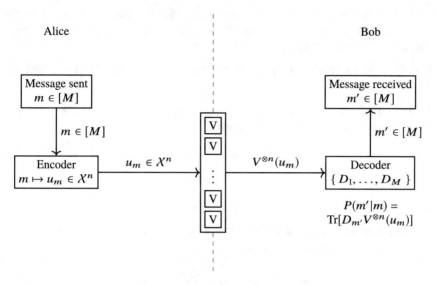

Fig. 4.4 Coding scheme for classical message transmission over n uses of the DMCQC V

$$e(\mathcal{C}, V^{\otimes n}) := \max_{m \in [M]} \text{Tr}\left[D_m^c V^{\otimes n}(u_m)\right] \quad \text{(maximal transmission error)}.$$
$$(4.32)$$

The quantities which we aim to maximize are for given transmission error $0 < \lambda < 1$, the maximal size of message sets for each blocklength n which allow the transmission error to be at most λ. We define the following quantities accordingly.

$$\overline{N}(V, n, \lambda) := \max\{M \in \mathbb{N} : \exists \, (n, M) \text{ code } \mathcal{C} \text{ with } \overline{e}(\mathcal{C}, V^{\otimes n}) \le \lambda\}, \quad (4.33)$$

$$N(V, n, \lambda) := \max\{M \in \mathbb{N} : \exists \, (n, M) \text{ code } \mathcal{C} \text{ with } e(\mathcal{C}, W^{\otimes n}) \le \lambda\}. \quad (4.34)$$

The quantities defined above generally grow exponentially with n (except when the channel is completely useless for message transmission). We will determine the asymptotic behavior of the *transmission rates*

$$\frac{1}{n} \log N(V, n, \lambda), \quad \text{and} \quad \frac{1}{n} \log \overline{N}(V, n, \lambda). \quad (4.35)$$

Exercise 4.1 Show that for each $n \in \mathbb{N}, \lambda \in (0, 1)$ and each cq channel V, the inequality

$$N(V, n, \lambda) \le \overline{N}(V, n, \lambda) \le \frac{1}{1 - \sqrt{\lambda}} N(V, n, \sqrt{\lambda}) \quad (4.36)$$

holds. The left inequality follows directly from the definitions. The proof for the right inequality directly carries over from a corresponding relation for classical discrete memoryless channels.

For characterizing the message transmission capacity of the DMCQC we need the following function.

Definition 4.8 (Holevo Quantity) Let $V : X \to S(\mathcal{H})$ be a cq channel and $q \in \mathcal{P}(X)$ a probability distribution. The function

$$\chi(q, V) := S(\overline{V}_q) - \sum_{x \in X} q(x) S(V(x)) \tag{4.37}$$

with $\overline{V}_q := \sum_{x \in X} q(x) V(x)$ is called the *Holevo quantity* of (q, V).

A convenient equivalent of the above expression can be given in terms of the quantum relative entropy. It holds by definition

$$\chi(q, V) = \sum_{x \in X} q(x) \, D(V(x) \| \overline{V}_q). \tag{4.38}$$

Moreover, the *Holevo Information* of the channel V is defined by

$$C(V) := \sup_{p \in \mathcal{P}(X)} \chi(p, V). \tag{4.39}$$

Theorem 4.5 (Coding Theorem and Converse) *Let* $V : X \to S(\mathcal{H})$ *be a cq channel. The following statements are true.*

1. $\forall \lambda > 0 : \liminf_{n \to \infty} \frac{1}{n} \log N(V, n, \lambda) \geq C(V)$.
2. $\inf_{\lambda > 0} \limsup_{n \to \infty} \frac{1}{n} \log \overline{N}(V, n, \lambda) \leq C(V)$.

The first statement in the above theorem is usually called the *coding theorem* for the discrete memoryless classical-quantum channels, and the second claim the *(weak) converse* to the coding theorem.

   Theorem 4.5 determines the *message transmission capacity* of a DMCQC V by $C(V)$. The complete theorem was proved by Holevo as well as Schumacher and Westmoreland in 1998 and 1997 respectively, many years after Holevo originally proved bounds on the accessible information of the channel in 1973.

   In fact, the second claim above can be replaced by the stronger statement

2'. $\forall \lambda > 0 : \limsup_{n \to \infty} \frac{1}{n} \log \overline{N}(V, n, \lambda) \leq \sup_{p \in \mathcal{P}(X)} \chi(p, V)$,

which is usually called the *strong converse* to the coding theorem for the DMCQC. The claim in 2'. also holds, but we will not give a proof of this statement here.

4.4.2 The Discrete Memoryless Quantum Channel

In this section, we consider the task of classical message transmission for memoryless quantum channels, which take quantum systems as inputs (and output quantum systems). The noise characteristics of such a channel are completely described by a c.p.t.p. map (which explains why such maps are also-called *quantum channels*).

Definition 4.9 Let $\mathcal{N} \in \mathcal{C}(\mathcal{H}, \mathcal{K})$ be a c.p.t.p. map. The *discrete memoryless quantum channel (DMQC)* generated by $\mathcal{N}$ is the family

$$\{\mathcal{N}^{\otimes n} : n \in \mathbb{N}\}. \tag{4.40}$$

The transmission map for n uses of the DMQC $\mathcal{N}$ is $\mathcal{N}^{\otimes n}$.

Next we define message transmission codes for DMQCs.

Definition 4.10 An (n, M)-code for transmission of classical messages over the DMQC $\mathcal{N} \in \mathcal{C}(\mathcal{H}, \mathcal{K})$ is a family $\mathcal{C} = (W(m), D_m)_{m=1}^M$ with

$$W(m) \in \mathcal{S}(\mathcal{H}^{\otimes n}), \tag{4.41}$$

and

$$D_m \in [0, \mathbb{1}_{\mathcal{K}}^{\otimes n}] \text{ for all } m \in [M] \text{ and } \sum_{m=1}^M D_m = \mathbb{1}_{\mathcal{K}}^{\otimes n}. \tag{4.42}$$

We define the *average transmission error* by

$$\bar{e}(\mathcal{C}, \mathcal{N}^{\otimes n}) := \frac{1}{M} \sum_{m=1}^M \text{Tr}\left[D_m^c \mathcal{N}^{\otimes n}(W(m))\right], \tag{4.43}$$

where we again use the notation $D_m^c := \mathbb{1}_{\mathcal{H}}^{\otimes n} - D_m$ for each $m \in [M]$.

As in the case of memoryless classical-quantum channels, we will determine the optimal asymptotical rates for classical message transmission. Therefore, we define for a given c.p.t.p. map $\mathcal{N}$ and each $\lambda \in [0, 1], n \in \mathbb{N}$

$$\overline{N}(\mathcal{N}, n, \lambda) := \max \left\{ M : \exists (n, M) - \text{message transmission code } \mathcal{C} \atop \text{with } \bar{e}(\mathcal{C}, \mathcal{N}^{\otimes n}) \leq \lambda \right\}. \tag{4.44}$$

Exercise 4.2 (Maximal Error) In the above definition, we only defined the average transmission error. Define a corresponding maximal error function and optimal message set sizes $N(\mathcal{N}, n, \lambda)$. Show that

$$(1 - \lambda^2)\overline{N}(\mathcal{N}, n, \lambda^2) \; \leq \; N(\mathcal{N}, n, \lambda) \; \leq \; \overline{N}(\mathcal{N}, n, \lambda) \qquad (4.45)$$

holds.

The rightmost inequality is clear operationally as a message set that satisfies an upper bound on the maximal error, also satisfies the same upper bound on the average error. The leftmost inequality is due to Markov's inequality, between the average and probability of maximal error.

Exercise 4.3 Show, that for each $\lambda \in (0, 1)$

$$\overline{N}(\mathcal{N}, m, \lambda) \leq \overline{N}(\mathcal{N}, n, \lambda) \qquad (4.46)$$

holds, if $m \leq n$.

This statement becomes clear when one considers the fact that more uses of the channel can only increase the number of communicated messages. Let $\mathcal{N} \in \mathcal{C}(\mathcal{H}, \mathcal{K})$. Define for each $k \in \mathbb{N}$,

$$C^{(1)}(\mathcal{N}^{\otimes k}) \; := \; \sup\{\chi(p, \mathcal{M} \circ V) : |\mathcal{Y}| < \infty, \; p \in \mathcal{P}(\mathcal{Y}), \; V : \mathcal{Y} \to \mathcal{S}(\mathcal{H}^{\otimes k})\}.$$

$$(4.47)$$

We set

$$C(\mathcal{N}) := \sup_{k \in \mathbb{N}} \frac{C^{(k)}(\mathcal{N}^{\otimes k})}{k}. \qquad (4.48)$$

Next we state the coding theorem and converse for classical message transmission over quantum discrete memoryless channels.

Theorem 4.6 *Let* $\mathcal{N} \in \mathcal{C}(\mathcal{H}, \mathcal{K})$. *It holds*

1. $\forall \lambda > 0 : \liminf\limits_{n \to \infty} \frac{1}{n} \log \overline{N}(\mathcal{N}, n, \lambda) \; \geq \; C(\mathcal{N})$.
2. $\inf\limits_{\lambda > 0} \limsup\limits_{n \to \infty} \frac{1}{n} \log \overline{N}(\mathcal{N}, n, \lambda) \; \leq \; C(\mathcal{N})$.

   Compared to the Holevo information $C(W)$ for a discrete memoryless cq channel defined in Eq. (4.39), the above-defined capacity function $C(\mathcal{N})$ for a quantum DMC is of a more complex structure. While $C(W)$ can be evaluated just by a maximization problem for the generic cq channel W, $C(\mathcal{N})$ is a so-called *multi-letter formula* which means that in principle one has to solve a separate optimization problem for each instance $\mathcal{N}^{\otimes k}$, $k \in \mathbb{N}$. For quite some time it was a major open question whether or not the problem could be reduced such that $C(\mathcal{N}) = C^{(1)}(\mathcal{N})$ holds, but eventually an example of a channel $\mathcal{N}$ with $C(\mathcal{N}) > \mathcal{C}^{(1)}(\mathcal{N})$ was given. However, a discussion of the example is beyond the scope of this book. The expression for $C(\mathcal{N})$ can be simplified a bit, i.e. the supremum in Eq. (4.48) can be replaced by a limit.

Proposition 4.4 *Let* $\mathcal{N} \in \mathcal{C}(\mathcal{H}, \mathcal{K})$. *It holds*

$$C(\mathcal{N}) = \lim_{k \in \mathbb{N}} \frac{C^{(k)}(\mathcal{N}^{\otimes k})}{k}.$$
(4.49)

4.4.3 Some Properties of the Holevo Quantity

So far we have assigned an information theoretic meaning to the Holevo quantity. Here, we mention some of its properties.

Proposition 4.5 (Additivity) *Let* $V : \mathcal{X} \to \mathcal{S}(\mathcal{K})$, $W : \mathcal{Y} \to \mathcal{S}(\mathcal{H})$ *be classical-quantum channels. The following claims are true*

1. $\sup_{p \in \mathcal{P}(\mathcal{X} \times \mathcal{Y})} \chi(p, V \otimes W) = \sup_{q \in \mathcal{P}(\mathcal{X})} \chi(q, V) + \sup_{r \in \mathcal{P}(\mathcal{Y})} \chi(r, W)$.
2. For each $n \in \mathbb{N}$, *it holds*

$$\frac{1}{n} \sup_{p \in \mathcal{P}(\mathcal{X}^n)} \chi(p, V^{\otimes n}) = \sup_{q \in \mathcal{P}(\mathcal{X})} \chi(q, V).$$
(4.50)

Proposition 4.6 (Data Processing) *Let* $V : \mathcal{X} \to \mathcal{S}(\mathcal{K})$, $p \in \mathcal{P}(\mathcal{X})$. *For each quantum channel* $\mathcal{N} \in \mathcal{C}(\mathcal{K}, \mathcal{H})$, *it holds*

$$\chi(p, V) \geq \chi(p, \mathcal{N} \circ V) \qquad \text{(Data processing inequality)}.$$
(4.51)

4.5 Entanglement-Assisted Classical Communication

In the preceding sections, we have discussed the task of classical message transmission over discrete memoryless quantum channels. We noticed that we end up with a so-called multi-letter formula for the corresponding capacity, which turns out to be very unsatisfactory when it comes to its calculation.

A very interesting scenario arises if the sender and receiver have access to shared entanglement in addition to the quantum channel. In case enough entanglement is present, the corresponding capacity turns out to be described by a single-letter formula, as we will see in this section. The capacity formula we derive shows that sometimes entanglement helps achieve higher classical message transmission capacities for channel transmission. Enhancement of communication capacity under different channel models (multiple access channels) has also been considered in [Nö19].

The main goal of this section is to give a characterization of the optimal asymptotic classical message transmission rates over a DMQ channel if the sender and receiver can choose an arbitrary pure entangled state to assist coding.

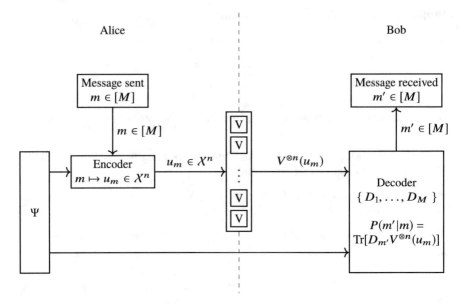

Fig. 4.5 Entanglement assisted channel coding. Alice and Bob share some pre-shared entanglement which we can model as a fixed state independent of the message (Ψ) of which one side is communicated without error to Bob. Alice then uses her part of the pre-shared entanglement to encode the message that goes through the quantum channel. Bob decodes using a global measurement on the output of the channel and his part of the pre-shared entanglement

The general coding scheme for entanglement-assisted classical message transmission over n uses of a DMQC $\mathcal{N}$ with assistance of a shared pure state $\Psi = |\psi\rangle\langle\psi|$ is depicted below.

We aim to determine the optimal asymptotically achievable message transmission rates in the above scenario. First we give the precise definition for *entanglement-assisted* (EA). The structure of the entanglement-assisted codes is displayed in Fig. 4.5.

Definition 4.11 Let $\mathcal{N} \in \mathcal{C}(\mathcal{H}_A \otimes \mathcal{H}_B)$ be a c.p.t.p. map. An (n, M)-*EA message transmission code* for $\mathcal{N}$ is a family $\mathcal{C} := (\Psi, \mathcal{E}_m, D_m)_{m=1}^M$ where

- $\Psi := |\psi\rangle\langle\psi| \in \mathcal{S}(\mathcal{K}_A \otimes \mathcal{K}_B)$ is a pure state shared by sender and receiver,
- $\mathcal{E}_m \in \mathcal{C}(\mathcal{K}_A, \mathcal{H}_A^{\otimes n})$ is a c.p.t.p. map for each $m \in [M]$, and
- $D_m \in \mathcal{L}(\mathcal{H}_B^{\otimes n}, \mathcal{K}_B)$ is a matrix, such that $0 \leq D_m \leq \mathbb{1}_{\mathcal{H}_B^{\otimes n} \otimes \mathcal{K}_B}$, and $\sum_{m=1}^M D_m = \mathbb{1}_{\mathcal{H}_B^{\otimes n} \otimes \mathcal{K}_B}$.

The *average error* of the code $\mathcal{C}$ is defined by

$$\bar{e}_{EA}(\mathcal{C}, \mathcal{N}^{\otimes n}) := \frac{1}{M} \sum_{m=1}^M \mathrm{Tr}\left[D_m^c((\mathcal{N}^{\otimes n} \circ \mathcal{E}_m) \otimes \mathrm{id}_{\mathcal{K}_B}(\Psi))\right], \qquad (4.52)$$

where we define, with some abuse of notation, $A^c := \mathbb{1} - A^c$ for each matrix A. As we did in the case of classical message transmission without entanglement assistance, we define

$$
\overline{N}_{EA}(\mathcal{N}, n, \epsilon) := \max \left\{ M : \begin{array}{l} \exists\, (n, M)\text{- EA message transm. code } \mathcal{C} \\[6pt] \text{with } \overline{e}_{EA}(\mathcal{C}, \mathcal{N}^{\otimes n}) \leq \epsilon \end{array} \right\} \quad (4.53)
$$

for each $\epsilon \in [0, 1]$ and $n \in \mathbb{N}$. To state the corresponding capacity theorem, we introduce another quantum entropic quantity.

Definition 4.12 (Quantum Mutual Information) For a c.p.t.p. map $\mathcal{N} \in \mathcal{C}(\mathcal{H}_A, \mathcal{H}_B)$ and a state $\rho \in \mathcal{S}(\mathcal{H}_A)$, the *quantum mutual information* is defined by

$$
I(\rho, \mathcal{N}) = S(\rho) + S(\mathcal{N}(\rho)) - S(\mathcal{N} \otimes \mathrm{id}_{\mathcal{K}}(|\psi\rangle \langle\psi|)), \quad (4.54)
$$

where ψ is the state vector of any purification of ρ.

Remark 4.2 Notice that the term on the right hand side of Eq. (4.54) above is indeed independent of the choice of purification.

Remark 4.3 Given the state $\omega_{AB} \in \mathcal{S}(\mathcal{H}_A \otimes \mathcal{H}_B)$, a closely related quantity, namely the mutual information of the state, is given by

$$
I(A; B, \omega) := S(A, \omega) + S(B, \omega) - S(AB, \omega), \quad (4.55)
$$

where $S(\gamma, \omega)$ indicates the von Neumann entropy of the marginal state ω_γ. The quantum mutual information of the state is closely related to the Holevo quantity (Eq. (4.48)). Consider the ensemble $\{p(x), \omega^x_{AB}\}$ with $\omega^x_{AB} \in \mathcal{S}(\mathcal{H}_A \otimes \mathcal{H}_B)$ and $p \in \mathcal{P}(\mathcal{X})$. We can define a classical-quantum (cq) state $\omega_{XAB} \in \mathcal{S}(\mathbb{C}^{|\mathcal{X}|} \otimes \mathcal{H}_A \otimes \mathcal{H}_B$, given some ONB $\{e_x\}_{x \in \mathcal{X}} \in \mathbb{C}^{|\mathcal{X}|}$, as

$$
\omega_{XAB} := \sum_{x \in \mathcal{X}} p(x) |e_x\rangle \langle e_x|^X \otimes \omega^x_{AB}. \quad (4.56)
$$

Note that we have used the suffix X to label the Hilbert space corresponding to alphabet $\mathcal{X}$. The conditional mutual information is then defined by

$$
I(A; B|X, \omega_{XB}) := \sum_{x \in \mathcal{X}} I(A; B, \omega^x_{AB}). \quad (4.57)
$$

Theorem 4.7 (Entanglement-Assisted Capacity) *Let* $\mathcal{N} \in \mathcal{C}(\mathcal{H}_A, \mathcal{H}_B)$. *It holds*

1. $\forall \epsilon > 0 : \liminf_{n \to \infty} \frac{1}{n} \log \overline{N}_{EA}(\mathcal{N}, n, \epsilon) \geq \sup_{\rho \in \mathcal{S}(\mathcal{H}_A)} I(\rho, \mathcal{N})$, *and*

2. $\inf_{\epsilon>0}$: $\limsup_{n\to\infty} \frac{1}{n} \log \overline{N}_{EA}(\mathcal{N}, n, \epsilon) \leq \sup_{\rho\in\mathcal{S}(\mathcal{H}_A)} I(\rho, \mathcal{N})$.

Exercise 4.4 Convince yourself that the EA (entanglement assisted) message transmission capacity is the same when the maximal error criterion is taken into account, instead of the average error.

The claims of Theorem 4.7 determine the input-state maximized quantum mutual information as the *entanglement-assisted classical capacity* of the QDMC $\mathcal{N}$. The following proposition states the existence of codes sufficient for proving Theorem 4.7 Item 1.

Proposition 4.7 *Let* $\mathcal{N} \in \mathcal{C}(\mathcal{H}_A, \mathcal{H}_B)$ *be a c.p.t.p. map, and* $\sigma \in \mathcal{S}(\mathcal{H}_A)$. *For each* $\epsilon > 0, \delta > 0$ *exists a number* n_0 *such that for each* $n > n_0$

$$\overline{N}_{EA}(\mathcal{N}, n, \epsilon) \geq \exp\left(n(I(\sigma, \mathcal{N}) - \delta)\right). \tag{4.58}$$

The strategy to prove the above claim will be to combine a number of instances of the channel, a given pure bipartite state and certain encoding maps to form an *effective* classical-quantum channel. It turns out that the classical message transmission codes for this cq channel can be reformulated to give EA message transmission codes for the original channel. To support this strategy, we need the following lemma.

Lemma 4.4 *Let* $\mathcal{H}$ *be a Hilbert space,* $\dim \mathcal{H} := d$, $\sigma \in \mathcal{S}(\mathcal{H})$, *and*

$$\psi = \sum_{i=1}^{d} \sqrt{\alpha_i} v_i \otimes v_i \tag{4.59}$$

be a Schmidt decomposition of a purification ψ *of* σ ($\alpha_i = 0$ *may occur for some* i), *and* $k \in \mathbb{N}$. *There is a family*

$$\{\tilde{\mathcal{E}}_x\}_{x\in\mathcal{X}} \subset \mathcal{C}(\mathcal{H}^{\otimes k}, \mathcal{H}^{\otimes k}), \tag{4.60}$$

such that for each Hilbert space $\mathcal{K}$ *and each* $\mathcal{N} \in \mathcal{C}(\mathcal{H}, \mathcal{K})$ *with the cq channel* $V : \mathcal{X} \to \mathcal{S}(\mathcal{K}^{\otimes n} \otimes \mathcal{K}^{\otimes n})$,

$$V(x) := \mathcal{N}^{\otimes k} \circ \tilde{\mathcal{E}}_x \otimes \mathrm{id}_{\mathcal{H}}^{\otimes k}(|\psi\rangle\langle\psi|^{\otimes k}) \qquad (x \in \mathcal{X}). \tag{4.61}$$

The inequality

$$|k \cdot I(\rho, \mathcal{N}) - \chi(q_*, V)| \leq 2d \cdot \log(k + 1) \tag{4.62}$$

is fulfilled with q_* *being the equidistribution on* $\mathcal{X}$.

Lemma 4.5 *Let* $\mathcal{N} \in \mathcal{C}(\mathcal{H}_A, \mathcal{H}_B)$, $\Psi \in \mathcal{S}(\mathcal{H}_A \otimes \mathcal{K}_B)$ *be a pure state,* $\{\mathcal{E}_m\}_{m=1}^{M} \subset \mathcal{C}(\mathcal{H}_A, \mathcal{H}_A)$, *and* $q \in \mathcal{P}([M])$. *Define the c.p.t.p. map* $\mathcal{E}(\cdot) := \sum_{m\in[M]} q(m)\mathcal{E}_m(\cdot)$, and $\rho_A = \text{Tr}_{\mathcal{K}_B} \Psi$. With the cq channel $V : [M] \to S(\mathcal{H}_B \otimes \mathcal{K}_B)$,

$$m \mapsto V(m) := \mathcal{N} \circ \mathcal{E}_m \otimes \text{id}_{\mathcal{K}_B}(\Psi), \qquad (4.63)$$

it holds

$$\chi(q, V) \leq I(\mathcal{E}(\rho_A), \mathcal{N}). \qquad (4.64)$$

We are now equipped with the prerequisites necessary to prove Theorem 4.7. The interested reader may consult [Jan20] for the full proof and mathematical treatment of the problem.

4.6 Information-Theoretic Security and CQQ Wiretap Model

An important aspect in information theory is security. Information-theoretic security, modeled by the *wiretap* channel that connects the sender to two receivers, one legal and the other wiretapper, is performed using a stochastic encoding procedure. The encoder first uses random codes designed for message transmission Sect. 4.4.1, and then uses part of these codes to confuse the wiretapper. This procedure known as equivocation, makes sure that the outcome of the channel at the wiretapper's end is arbitrarily close to a fixed state, independent of the encoding. We will see that this gives a positive rate of secure messages transmitted to the legal receiver, in case the channel connecting the legal parties is better (less noisy) than the one between the sender and the wiretapper.

Wyner [Wyn75] introduced the classical wiretap channel, and considered a subclass of channels known as the degraded wiretap channels, before Csiszár and Körner [CK78] addressed the general case. As mentioned above, the model can be described by two channels from the sender (Alice) to the legal receiver (Bob) and to the eavesdropper (Eve), respectively. In transmission theory the goal is to send messages to the legal receiver, while the wiretapper is to be kept ignorant. The wiretap channel was generalized to the setting of quantum information theory in [CWY04, Dev05]. Formally, in contrast to the classical case, quantum-mechanically the channel can be described by a single quantum operation T (see Theorem 2.6), from Alice to the joint system of Bob and Eve together: then we can define the legal channel $W = \text{Tr}_B \circ T$ and the wiretapper channel $V = \text{Tr}_E \circ T$. Note that (unlike the classical case) this pair of channels cannot be arbitrary. This has to do with the no-cloning theorem: Alice's input state cannot be duplicated and then sent through both channels.

However, here we will restrict ourselves to the cq-channel case, where Alice's input is described by a letter $x \in X$ from a finite alphabet. Then we can define the classical-quantum wiretap channel in a simple way.

Definition 4.13 A *wiretap classical-quantum-quantum (cqq) channel* is a pair (W, V) of two discrete memoryless cq-channels $W : \mathfrak{X} \longrightarrow \mathcal{S}(\mathcal{H}_B)$ and $V : \mathfrak{X} \longrightarrow \mathcal{S}(\mathcal{E})$. When Alice sends a classical input $x^n \in \mathfrak{X}^n$, Bob (legal receiver) and Eve (eavesdropper) receive the states $W^{\otimes n}(x^n)$ and $V^{\otimes n}(x^n)$, respectively.

Here we adopt the additional notation of letting probability distributions be inputs to the channel, so that if P is a probability distribution on $\mathfrak{X}$ and V is a classical-quantum channel then

$$V(P) = \sum_{x \in \mathfrak{X}} P(x) V(x) \qquad (4.65)$$

which we use in the next definition. Indeed, when security is involved, it is necessary to have randomization at the encoder.

Definition 4.14 An (n, M, λ, μ)-*wiretap code* for the wiretap cqq-channel (W, V) is a collection $\{(P_i, D_i) : i \in [M]\}$ of pairs consisting of probability distributions P_i on $\mathfrak{X}^n$ and a POVM $(D_i)_{i=1}^N$ on $\mathcal{H}_B^{\otimes n}$ such that

$$\forall i \in [M] \quad 1 - \mathrm{Tr}\left[W^{\otimes n}(P_i) \cdot D_i\right] \leq \lambda, \qquad (4.66)$$

$$\forall i, j \in [M] \quad \frac{1}{2}\|V^{\otimes n}(P_i) - V^{\otimes n}(P_j)\|_1 \leq \mu. \qquad (4.67)$$

The largest M such that an (n, M, λ, μ)-wiretap code exists is denoted $M(n, \lambda, \mu)$. The *secure capacity* of (W, V) is then defined as

$$C_S(W, V) = \inf_{\lambda, \mu > 0} \liminf_{n \to \infty} \frac{1}{n} \log M(n, \lambda, \mu). \qquad (4.68)$$

Note that by the Fannes inequality [Fan73, Win16], the second condition (security) implies that for any random variable J taking values in $[M]$, $I(J; E^n) \leq \mu n \log |E| + h(\mu)$. It turns out that the right hand side can be made arbitrarily small while achieving the capacity, because μ as well as λ can be made to converge to 0 to any polynomial order.

Theorem 4.8 ([CWY04]) *The secure capacity of a wiretap cqq-channel is given by*

$$C_S(W, V) = \lim_{n \to \infty} \max_{U \to X^n \to B^n E^n} \frac{1}{n} \left(I(U; B^n, \omega_W) - I(U; E^n, \omega_V)\right), \qquad (4.69)$$

where the maximum is taken over all random variables satisfying the Markov chain relationships $U \to X^n \to B^n E^n$. The quantities are evaluated on

$$\omega_T := \sum_{(u^n, x^n) \in \mathcal{U}^n \times \mathfrak{X}^n} p(u^n, x^n) |u^n\rangle \langle u^n| \otimes T^{\otimes n}(x^n), \qquad (4.70)$$

with $p \in \mathcal{P}(\mathcal{U} \times \mathcal{X})$ *a probability distribution, generating random variables* U, X *given the Markov chain.*

Thus in the case of transmission theory, we have a positive secure capacity C_S when the channel parameters of the legal channel are better than those of the non-legal channel. The main step applied on message transmission codes that allows security is what is known as the *covering Lemma*, which shows that a stochastic encoder, by creating an equidistribution on part of the input codes, can puzzle the eavesdropper. We will see that this idea is used towards information theoretic security under a more general channel model known as the broadcast channel.

4.7   Public and Secure Identification

In this section, we consider the task of *message identification* over cq channels and cqq *wiretap channels*, whereby information theoretic security is modeled (see Sect. 4.6). The task of identification was first introduced by Ahlswede and Dueck [AD89b], realizing that while Shannon's theory of message transmission presumes that the receiver wants to know everything about the message, in reality he may be interested only in certain aspects of it. In other words, the receiver may want to compute a function of the message. The most extreme case is that of identification: for sent message m and an arbitrary message m', the receiver would like to be able to answer the question *Is* $m = m'$? as accurately as possible. Here, we study the identification capacity of cq channels with and without security constraints. In Sect. 4.7.1, we start by introducing the task without security constraints. Finally, in Sect. 4.7.2 we consider its capacity for secure identification.

4.7.1   Identification via CQ Channels

We considered the ability of the cq channel $W : \mathcal{X} \rightarrow \mathcal{S}(\mathcal{H}_B)$ for message transmission in Sect. 4.4.1 and quantified it by $C(W)$ defined in Eq. (4.39). Compared to message transmission, in identification theory we change the goal for Bob: We assume that he *only* wants to know if the transmitted message is equal to some j.

Definition 4.15 An $(n, N, \lambda_1, \lambda_2)$ *ID-code* is a set of pairs $\{(P_i, D_i) : i \in [N]\}$ where the P_i are probability distributions on $\mathcal{X}^n$ and the D_i are POVM elements, i.e. $0 \leq D_i \leq 1$, acting on $\mathcal{H}_B^{\otimes n}$, that $\forall i \neq j \in [N]$

$$\mathrm{Tr}\left[W^{\otimes n}(P_i) \cdot D_i\right] \geq 1 - \lambda_1 \tag{4.71}$$

and

$$\mathrm{Tr}\left[W^{\otimes n}(P_i) \cdot D_j\right] \leq \lambda_2. \tag{4.72}$$

The largest size N of an $(n, N, \lambda_1, \lambda_2)$ ID-code is denoted $N(n, \lambda_1, \lambda_2)$.

Using a stochastic encoder is essential in the theory of identification. The definition of an ID-code only partially fits the definition of a classical identification code in the following sense. There are applications of classical identification codes, where one assumes that there are several receivers, each only interested in one message, and all wanting to decide individually if *their* message was sent. The example given in [AD89b] is that of N sailors on a ship, and each sailor is related to one relative. On a stormy night, one sailor drowns in the ocean. One could now broadcast the name of the sailor to all relatives. However, this takes $\lceil \log_2 N \rceil$ bits. And the news is of interest only to one relative. If we now allow a certain error probability, we can broadcast an identification code using only $O(\log_2 \log_2 N)$ bits.

The ID-code for a quantum channel has the property that the received state cannot be used in general to ask for two different messages. The reason is that the POVMs $(D_i, \mathbb{1} - D_i)$ are in general not compatible. Therefore the realization of applications with more than one receiver, like in the example above, is not possible with an ID-code as defined. There are, however, applications where we have only two parties who want to check if they have the same text (such as watermarking, or in the communication complexity setting). Löber [L99] defined simultaneous ID-codes to overcome this limitation. In this code model, there has to be one single measurement which allows us to identify every message at the same time.

Definition 4.16 An ID-code $\{(P_i, D_i) : i \in [N]\}$ is called *simultaneous* if there is a POVM $(E_y)_{y \in \mathcal{Y}}$ acting on $B^{\otimes n}$ and subsets $\mathcal{A}_i \subseteq \mathcal{Y}$, such that $D_i = \sum_{y \in \mathcal{A}_i} E_y$ for all $i \in [N]$. The largest size of a simultaneous $(n, \lambda_1, \lambda_2)$ ID-code is denoted $N_{sim}(n, \lambda_1, \lambda_2)$.

In this case the measurement gives as a result some $y \in \mathcal{Y}$, and receiver i has to check whether $y \in \mathcal{A}_i$. Note that the definition can be expressed equivalently by requiring that the measurements $(D_1, \mathbb{1} - D_i)$ are all compatible, because this requires that there exists a common refinement of them, i.e. a POVM of which all $(D_1, \mathbb{1} - D_i)$ are coarse grainings.

Remark 4.4 If the D_i are not compatible, there is no way of measuring them all together jointly, but this does not mean that we have to give up. To identify a set of messages $i_1, \ldots, i_k$, we could simply apply the decoding POVMs $(D_{i_\kappa}, \mathbb{1} - D_{i_\kappa})$ sequentially in some order. That this is not a bad idea follows from the gentle measurement lemma [Win99]: since each measurement has a high probability of giving the correct outcome, the state is disturbed, but only *a little* in trace norm, so we can subject the next measurement as if nothing had happened at all.

The best analysis of this approach is using Sen's non-commutative union bound [Sen12] for general POVMs [Wil13]. Using this bound, we can see that if we have any ID-code with errors $\lambda_1, \lambda_2 \leq \lambda$, then we can correctly identify any set of $k \leq \frac{\epsilon^2}{4\lambda}$ messages, with error probability bounded by ϵ. This will not include all messages, since for the rates below the capacity, the error λ can be made to vanish exponentially and we get at least an exponentially large k.

Here we consider the identification capacity of a cq-channel, of which we distinguish a priori simultaneous and non-simultaneous flavors, following Löber [L99]:

Definition 4.17 The *(simultaneous) classical ID-capacity* of a cq-channel W is defined as

$$C_{ID}(W) = \inf_{\lambda > 0} \liminf_{n \to \infty} \frac{1}{n} \log \log N(n, \lambda, \lambda), \tag{4.73}$$

$$C_{ID}^{sim}(W) = \inf_{\lambda > 0} \liminf_{n \to \infty} \frac{1}{n} \log \log N_{sim}(n, \lambda, \lambda), \tag{4.74}$$

respectively.

Löber [L99] showed that for cq-channels, the simultaneous classical ID capacity is equal to the transmission capacity. Furthermore, he showed that the strong converse holds for simultaneous ID-codes. Later, Ahlswede and Winter [AW02] extended the strong converse to non-simultaneous ID-codes.

Theorem 4.9 ([L99, AW02]) *For any cq-channel W,*

$$C_{ID}^{sim}(W) = C_{ID}(W) = C(W), \tag{4.75}$$

and the strong converse holds: for all $\lambda_1 + \lambda_2 < 1$,

$$\lim_{n \to \infty} \frac{1}{n} \log \log N(n, \lambda_1, \lambda_2)$$

$$= \lim_{n \to \infty} \frac{1}{n} \log \log N_{sim}(n, \lambda_1, \lambda_2) = C(W). \qquad \blacksquare \tag{4.76}$$

Ahlswede and Winter also considered the case of a general (quantum-quantum) channel, but the results are much less complete. It is not even clear if in the general case the simultaneous capacity and the non-simultaneous ID-capacity coincide. See the subsequent papers [HW12] and the review [Win13] for a presentation of the state of the art.

Definitions 4.15 and 4.16 address Freeman Dyson's critique on the status quo of experiments, measurements, and detectors in particle physics (in our setting of operational tasks). According to Dyson, experiments as currently conducted in particle physics can only answer very specific questions. Analogous to our model, this corresponds to identification codes Definition 4.15, and in particular the use of *message-dependent* measurements. In comparison, the simultaneous identification codes provide universal measurements so that the relevant questions can be answered by classical post-processing.

4.7.2   Secure Identification

In Sect. 4.6, we saw that we have a positive secure capacity C_S when the channel parameters of the legal channel are better than those of the non-legal channel. This means we pay a price in the form of a smaller rate for secure transmission. We will observe that in the case of identification, the situation is different.

Definition 4.18 An $(n, N, \lambda_1, \lambda_2, \mu)$ *wiretap ID-code* for the wiretap cqq-channel (W, V) is a set of pairs $\{(P_i, D_i) : i \in [N]\}$, where the P_i are probability distributions on $\mathcal{X}^n$, and the D_i, $0 \leq D_i \leq \mathbb{1}$ denote operators on $\mathcal{H}_B^{\otimes n}$, such that for all $i \neq j \in [N]$ and for all $0 \leq F \leq \mathbb{1}_{E^n}$,

$$\mathrm{Tr}\left[W^{\otimes n}(Q_i)D_i\right] \geq 1 - \lambda_1, \tag{4.77}$$

$$\mathrm{Tr}\left[W^{\otimes n}(Q_j)D_i\right] \leq \lambda_2, \tag{4.78}$$

$$\mathrm{Tr}\left[V^{\otimes n}(Q_j)F\right] + \mathrm{Tr}\left[V^{\otimes n}(Q_i)(\mathbb{1} - F)\right] \geq 1 - \mu. \tag{4.79}$$

If the POVMs $(D_i, \mathbb{1} - D_i)$ are all compatible, we call the code *simultaneous*, as in the cq-channel case.

In contrast to the transmission problem, the decoding sets for the identification problem are not necessarily disjoint.

Condition Eq. (4.79) enforces that the wiretapper cannot very well distinguish the output states $Q_i V^{\otimes n}$ of the different messages. Indeed, it is equivalent to

$$\mu \geq \max_{0 \leq F \leq \mathbb{1}} \mathrm{Tr}\left[(V^{\otimes n}(Q_j) - V^{\otimes n}(Q_i))F\right] \tag{4.80}$$

$$= \frac{1}{2}\|V^{\otimes n}(Q_j) - V^{\otimes n}(Q_i)\|_1, \tag{4.81}$$

which by Helstrom's theorem [Hel69, NC10] means that even if Eve somehow knows that the message can only be either i or j with equal probability, then her error probability for discriminating these two alternatives is at least $\frac{1}{2}(1 - \mu) \approx \frac{1}{2}$.

The maximum N for which a $(n, N, \lambda_1, \lambda_2, \mu)$ wiretap ID-code exists is denoted by $N(n, \lambda_1, \lambda_2, \mu)$. For simultaneous wiretap ID-codes we denote the maximum $N_{sim}(n, \lambda_1, \lambda_2, \mu)$. We then define the (simultaneous) secure identification capacity of the wiretap channel as

$$C_{SID}(W, V) \quad = \inf_{\lambda_1,\lambda_2,\mu>0} \liminf_{n\to\infty} \frac{1}{n} \log\log N(n, \lambda_1, \lambda_2, \mu), \tag{4.82}$$

$$C_{SID}^{sim}(W, V) \quad = \inf_{\lambda_1,\lambda_2,\mu>0} \liminf_{n\to\infty} \frac{1}{n} \log\log N_{sim}(n, \lambda_1, \lambda_2, \mu), \tag{4.83}$$

respectively.

In this section we consider the wiretap cqq-channel and derive a multi-letter formula for its secure identification capacity. The idea is similar to the classical case. We use a combination of two codes. For the converse we generalize inequalities of [AZ95] and [FVDG99].

Theorem 4.10 (Dichotomy Theorem) *Let $C(W)$ be the capacity of the cq-channel W and let $C_S(W, V)$ be the secure capacity of the wiretap cqq-channel. Then,*

$$C_{SID}(W, V) = C_{SID}^{sim}(W, V) \tag{4.84}$$

$$= \begin{cases} C(W) & \text{if } C_S(W, V) > 0, \\ 0 & \text{if } C_S(W, V) = 0. \end{cases} \tag{4.85}$$

We briefly explain the coding idea for the direct part. The identification code is constructed by means of two fundamental codes, following [AD89a].

Let $0 < \epsilon < C$ be fixed. We know that there is a $\delta > 0$ such that for sufficiently large n there is an $(n, M', \lambda(n))$-code $C' = \left\{ \left(u'_j, \mathcal{D}'_j \mid j \in [M'] \right) \right\}$ for the cq-channel with code size $M' = \lceil 2^{n(C(W)-\epsilon)} \rceil$ and by Theorem 4.8 an $(\lceil \sqrt{n} \rceil, M'', \lambda(\sqrt{n}), \mu(\sqrt{n}))$ wiretap code $C'' = \left\{ \left(u'_k, \mathcal{D}'_k \mid k \in [M''] \right) \right\}$ for the wiretap cqq-channel $(W; V)$ with code size $M'' = \lceil 2^{\epsilon \sqrt{n}} \rceil$. Alice and Bob first create shared randomness with the help of the code C' at a rate equal to the channel capacity. A code with an arbitrary small positive rate is then sufficient to use the method of Ahlswede and Dueck by sending and decoding the function values. For this purpose, the code C'' is used. Furthermore, it can be shown that if $C_{SID}(W, V) > 0$, then $C_S(W, V) > 0$.

4.8 Channel Uncertainty:[1] Compound and Arbitrarily Varying Models

In real world communication using either quantum or classical systems, the parameter determining the channel in use may belong to an uncertainty set, rendering the protocols that assume the channel to be perfectly known, practically obsolete. Given such uncertainty, when using the channel many times as is done in Shannon theoretic information processing tasks, assuming the channel to be memoryless or fully stationary is not realistic. In this part, we consider three models that include channel uncertainty without attempting to reduce it via techniques such as channel identification or tomography. We refer to these models as the compound, arbitrarily varying and fully quantum arbitrarily varying channel models. Each of these models

---

[1]Not to be confused with the other use of the word *uncertainty* in this book, namely that related to Heisenberg's uncertainty principle.

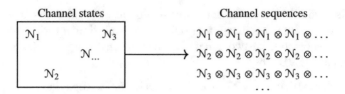

Fig. 4.6 Compound channel model

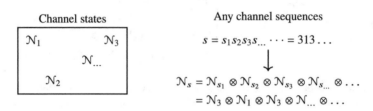

Fig. 4.7 Arbitrarily varying channel model

are considered here for transmission of entanglement, classical and secure messages simultaneously, between a sender and receiver.

Informally, the first two channel models considered here consist of a set of quantum channels $\{\mathcal{N}_s\}_{s \in S}$ known to the communicating parties. In the compound model (see Fig. 4.6), communication is done under the assumption that asymptotically, one of the channels from this set (unknown to the parties) is used in a memoryless fashion. The codes used in this model therefore have to be reliable for the whole family $\{\mathcal{N}_s^{\otimes l}\}_{s \in S}$ of memoryless channels for large enough values of $l \in \mathbb{N}$.

In the arbitrarily varying model (see Fig. 4.7), given a number of channel uses l, an adversarial party chooses the sequence $s^l = (s_1, \ldots, s_l) \in S^l$ unknown to the communication parties, to yield the channel $\mathcal{N}_{s^l} = \bigotimes_{i=1}^{l} \mathcal{N}_{s_i}$. The adversary may choose this sequence knowing the encoding procedure used by the sender. The code therefore has to be reliable for the whole family of memoryless channels $\{\mathcal{N}_{s^l}\}_{s^l \in S^l}$. Finally, in the third channel model, namely that of the fully quantum arbitrarily varying channel (see Fig. 4.8), the assumption of memoryless communication is dropped. Precise definition of this channel model is given in Sect. 4.8.2.5. As the power to choose the state of the channel in the arbitrarily varying models is ascribed to a jammer, we refer to the state of the channel chosen by the adversary as the jamming attack.

We consider attacks that are performed directly at the physical layer with the aim of disrupting the physical transmission itself. Such attacks can target a specific single user within the system, but also the overall system itself. Reliable communication between legitimate users is the indispensable basis for any information processing. In the worst case, the jammer is able to perform a denial-of-service (DoS) attack which means that no communication is possible at all. In [HB20a] it was shown that it is impossible to algorithmically detect such fundamental physical jamming attacks. The undetectability of DoS attacks has

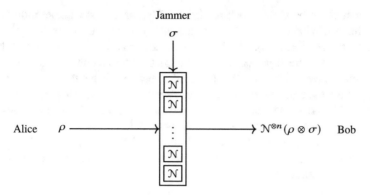

Fig. 4.8 Fully quantum arbitrarily varying channel model

crucial implications and consequences on higher layers of communication systems. It was discussed in [HB20a] that it is possible to obtain resilience by design by invoking additional resources to stabilize the communication directly at the physical layer. Techniques to achieve resilience by design have been analyzed in [ABBN13, HB14, HB19, HB20b].

In Sect. 4.8.2 based on results from [BJS19a], we consider an integrated task in which the communicating parties wish to transmit classical messages and entanglement under the channel models mentioned above. Precise definitions of the protocols will be given therein. Clearly, the resulting capacity-region achieving codes here will reduce to those appropriate for each of these two tasks, when only one dimension of the region is considered.

In Sect. 4.8.3, based on results from [BJS19b], we consider an integrated task where the communicating parties wish to establish secure and public communication. Here, our channel may be appropriately named a compound classical-quantum broadcast channel. Again our codes will achieve the channel's two dimensional capacity region that also contains the capacity of the channel for each task. In this section, we leave out the capacity under assumptions of the arbitrarily varying model, and instead delve deeper into the compound model. Information theoretically, the compound model has yielded intriguing properties. One of the interesting information theoretic properties of the compound channel is that in general, a strong converse cannot be established on the capacity of the compound channel for message transmission when upper-bounding of the average decoding error is considered. This holds even for finite uncertainty sets [Ahl67, AW69, BBJN13]. This observation implies that a second order capacity theorem cannot be developed in this case. Further, calculation of the so-called ϵ-capacity of the compound channel under the average error criterion is still an open question. We note however, that determining a second order ϵ-capacity for the compound channel is not possible, due to the observation that there are examples of the compound channel where the optimistic ϵ-capacity is strictly larger than its pessimistic one (see [BSP18] Remark 13).

In what could be viewed as a reduction of the broadcast channel introduced in Sect. 4.8.3, the wiretap model connects a sender with two receivers, a legal one and a wiretapper (see Sect. 4.6). The legitimate receiver accesses the output of the first channel and the wiretapper observes the output of the second channel. A code for the channel conveys information to the legal receiver such that the wiretapper knows nothing about the transmitted information. In Sects. 4.8.4 and 4.8.5, based on results from [BCCD14a] and [BDW19] respectively, we generalize the results of Sects. 4.6 and 4.7 to the mentioned channel uncertainty models.

Before stating any results, we include the following notational section to accompany the notation already introduced in this work, specifically for the needs of the present section.

4.8.1 Notations and Conventions

Here we introduce notations that have not yet been introduced in the quantum information theory part of this book. To each subspace $\mathcal{F} \subset \mathcal{H}$, we can associate a unique projection $q_{\mathcal{F}}$ whose range is the subspace $\mathcal{F}$, and we write $\pi_{\mathcal{F}}$ for the maximally mixed state on $\mathcal{F}$, i.e.

$$\pi_{\mathcal{F}} := \frac{q_{\mathcal{F}}}{\operatorname{Tr}(q_F)}. \tag{4.86}$$

$\mathcal{C}^{\downarrow}(\mathcal{H}_A, \mathcal{H}_B)$ stands for the set of completely positive trace non-increasing maps between $\mathcal{L}(\mathcal{H}_A)$ and $\mathcal{L}(\mathcal{H}_B)$. In what follows, $\mathcal{U}(\mathcal{H})$ will denote the group of unitary operators acting on $\mathcal{H}$. For a Hilbert space $\mathcal{G} \subset \mathcal{H}$, we will always identify $\mathcal{U}(\mathcal{G})$ with a subgroup of $\mathcal{U}(\mathcal{H})$. For any projection $q \in \mathcal{L}(\mathcal{H})$ we set $q^{\perp} := 1_{\mathcal{H}} - q$.

Each projection $q \in \mathcal{L}(\mathcal{H})$ defines a completely positive trace non-increasing map Q given by $Q(a) := qaq$ for all $a \in \mathcal{L}(\mathcal{H})$. In a similar fashion, any $U \in \mathcal{U}(\mathcal{H})$ defines a $\mathcal{U} \in \mathcal{C}(\mathcal{H}, \mathcal{H})$ by $\mathcal{U}(a) := UaU^{\dagger}$ for $a \in \mathcal{L}(\mathcal{H})$. The coherent information for $\mathcal{N} \in \mathcal{C}(\mathcal{H}_A, \mathcal{H}_B)$ and $\rho \in \mathcal{S}(\mathcal{H}_A)$ is defined by

$$I_c(\rho, \mathcal{N}) := S(\mathcal{N}(\rho)) - S((\mathrm{id}_{\mathcal{H}_A} \otimes \mathcal{N})(|\psi\rangle\langle\psi|)), \tag{4.87}$$

where $\psi \in \mathcal{H}_A \otimes \mathcal{H}_A$ is an arbitrary purification of the state ρ. A short-hand notation $S_e(\rho, \mathcal{N}) := S((\mathrm{id}_{\mathcal{H}_A} \otimes \mathcal{N})(|\psi\rangle\langle\psi|))$ to denote entropy exchange is also used in the literature. A useful equivalent definition of $I_c(\rho, \mathcal{N})$ is given in terms of $\mathcal{N} \in \mathcal{C}(\mathcal{H}_A, \mathcal{H}_B)$ and any complementary channel $\hat{\mathcal{N}} \in \mathcal{C}(\mathcal{H}_A, \mathcal{H}_e)$ where $\mathcal{H}_e$ denotes the Hilbert space of the environment. Due to Stinespring's dilation theorem (see Theorem 2.6), $\mathcal{N}$ can be represented as

$$\mathcal{N}(\rho) = \operatorname{Tr}_{\mathcal{H}_e}(v\rho v^*) \tag{4.88}$$

for $\rho \in \mathcal{S}(\mathcal{H}_A)$, where $v : \mathcal{H}_A \to \mathcal{H}_B \otimes \mathcal{H}_e$ is a linear isometry. The complementary channel $\hat{N} \in \mathcal{C}(\mathcal{H}_A, \mathcal{H}_e)$ of N is given by

$$\hat{N}(\rho) := \mathrm{Tr}_{\mathcal{H}_B}(v\rho v^*). \tag{4.89}$$

The coherent information can then be written as

$$I_c(\rho, N) = S(N(\rho)) - S(\hat{N}(\rho)). \tag{4.90}$$

This quantity can also be defined in terms of the bipartite state $\sigma \in \mathcal{S}(\mathcal{H}_A \otimes \mathcal{H}_B)$ with

$$\sigma := \mathrm{id}_{\mathcal{H}_A} \otimes N(|\psi\rangle \langle\psi|) \tag{4.91}$$

as

$$I(A\rangle B, \sigma) := S(\sigma^B) - S(\sigma), \tag{4.92}$$

where σ^B is the marginal state given by $\sigma^B := \mathrm{Tr}_A(\sigma)$ and we have the identity

$$I_c(\rho, N) = I(A\rangle B, \sigma). \tag{4.93}$$

For the approximation of arbitrary compound channels (introduced in the next section) by finite ones, we use the diamond norm $\| \cdot \|_\diamond$, given for any $N : \mathcal{L}(\mathcal{H}_A) \to \mathcal{L}(\mathcal{H}_B)$ by

$$\| N \|_\diamond := \sup_{n \in \mathbb{N}} \max_{a \in \mathcal{L}(\mathbb{C}^n \otimes \mathcal{H}), \|a\|_1 = 1} \| (\mathrm{id}_n \otimes N)(a) \|_1, \tag{4.94}$$

where $\mathrm{id}_n : \mathcal{L}(\mathbb{C}^n) \to \mathcal{L}(\mathbb{C}^n)$ is the identity channel. We state the following facts about $\| \cdot \|_\diamond$ (see e.g [KSV02]). First, $\|N\|_\diamond = 1$ for all $N \in \mathcal{C}(\mathcal{H}_A, \mathcal{H}_B)$. Thus, $\mathcal{C}(\mathcal{H}_A, \mathcal{H}_B) \subset S_\diamond$, where $S_\diamond$ denotes the unit sphere of the normed space $(\mathcal{L}(\mathcal{H}_A), \mathcal{L}(\mathcal{H}_B), \| \cdot \|_\diamond)$. Moreover, $\|N_1 \otimes N_2\|_\diamond = \|N_1\|_\diamond \|N_2\|_\diamond$ for arbitrary linear maps $N_1, N_2 : \mathcal{L}(\mathcal{H}_A) \to \mathcal{L}(\mathcal{H}_B)$. Throughout this section we have made use of the idea of nets to approximate arbitrary compound quantum channels using ones with finite uncertainty sets. This idea is presented in appendix section of [BJS19a].

In Sect. 4.8.3, the set of probability distributions on the finite alphabet $\mathcal{X}$ of cardinality $|\mathcal{X}|$ will be denoted by $\mathcal{P}(\mathcal{X})$. For $n \in \mathbb{N}$, we define $\mathcal{X}^n := \{(x_1, \ldots, x_n) : x_i \in \mathcal{X}, \forall i \in \{1, \ldots, n\}\}$. The sequence $\mathbf{x}$ will denote elements of $\mathcal{X}^n$. Also, we use bold letters to denote vectors (sequences with more that one element). The probability distribution $p^{\otimes n} \in \mathcal{P}(\mathcal{X}^n)$ will be given by the n-fold product of $p \in \mathcal{P}(\mathcal{X})$, namely $p^{\otimes n}(\mathbf{x}) = p(x_1) \ldots p(x_n)$ with $\mathbf{x} = (x_1, \ldots, x_n)$. For any number $M \in \mathbb{N}$, we use $[M] := \{1, \ldots, M\}$.

The classical quantum (cq) channel $W : \mathcal{X} \to \mathcal{S}(\mathcal{H})$ is a completely positive trace preserving map from alphabet $\mathcal{X}$ to the set of states on Hilbert space $\mathcal{H}$. We denote the set of all such maps by $CQ(\mathcal{X}, \mathcal{H})$. This set is equipped with the norm $\| \cdot \|_{CQ}$ defined for $W \in CQ(\mathcal{X}, \mathcal{H})$ by

$$\| W \|_{CQ} := \max_{x \in \mathcal{X}} \| W(x) \|_1, \tag{4.95}$$

where $\| \cdot \|_1$ is the trace norm on $\mathcal{L}(\mathcal{H})$. We use the term cqq channel for map $V \in CQ(\mathcal{X}, \mathcal{H}_1 \otimes \mathcal{H}_2)$ with two outcomes in two sets of states on two Hilbert spaces. With a slight abuse of notation, we write $a^c := \mathbb{1}_{\mathcal{H}} - a$ for $a \in \mathcal{L}(\mathcal{H})$.

4.8.2   Simultaneous Transmission of Classical and Quantum Information

In this section,[2] we consider universal codes for simultaneous transmission of classical messages and entanglement through quantum channels, possibly under attack by a malignant third party. These codes are robust to different kinds of channel uncertainty. To construct such universal codes, we invoke and generalize properties of random codes for classical and quantum message transmission through quantum channels. We show these codes to be optimal by giving a multi-letter characterization of regions corresponding to the capacity of compound quantum channels for simultaneously transmitting and generating entanglement with classical messages. Also, we give dichotomy statements in which we characterize the capacity of arbitrarily varying quantum channels for simultaneous transmission of classical messages and entanglement. These include cases where the malignant jammer present in the arbitrarily varying channel model is classical (chooses channel states of product form) and fully quantum (is capable of general attacks not necessarily of product form).

As considered in the first part of this work, the quantum channel has different capacities for information transmission. One may consider the capacity of the channel for public [Hol98, SW97] or secure [Dev05, CWY04] classical message transmission, entanglement transmission or entanglement generation [Dev05] to name a few. These communication scenarios have been considered subsequently under channel uncertainty [BBN09, Mos15, BGW17, BJK17, Ahl78, ABBN12]. Simultaneous transmission of classical and quantum messages, the subject of this section, has also been of interest [DS05]. This includes scenarios where the communication parties would like to enhance their classical message transmission by sharing quantum information primarily at their disposal or vice versa [BSST99, HW10a, HW10b]. The body of research in this area is clearly interesting, when regions beyond those achieved by simple time-sharing (Fig. 4.9) between

---

[2]This section is based on [BJS19a].

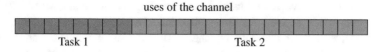

uses of the channel

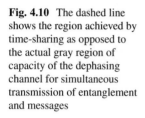

Fig. 4.9   Time-sharing strategy

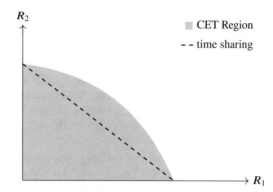

Fig. 4.10   The dashed line shows the region achieved by time-sharing as opposed to the actual gray region of capacity of the dephasing channel for simultaneous transmission of entanglement and messages

established classical message and quantum information transmission codes are reached.

Simultaneous transmission of classical messages and entanglement is a nontrivial problem even if capacity achieving codes for the corresponding univariate transmission goals are at hand. It was already observed in [DS05] for perfectly known quantum channels that the naive time sharing strategy is generally insufficient to achieve the full capacity region (Fig. 4.10). Examples of channels where coding beyond time-sharing is indispensable does not depend on constructing pathologies. They are readily found even within the standard arsenal of qubit quantum channels, e.g. the dephasing qubit channels [DS05].

We derive codes for simultaneous transmission of classical messages and entanglement that are robust to the three types of uncertainty mentioned above. The codes used here for the compound model are different from those used for the point to point communication in [DS05] when considering the special case of $|S| = 1$. Given that the input state approximation techniques used therein prove insufficient in the presence of channel state uncertainty, in the present section we use the decoupling approach first established in [Kle07]. We combine robust random codes for classical message transmission from [Mos15] and a generalization of (decoupling based) entanglement transmission codes from [BBN09] to construct appropriate simultaneous codes for compound quantum channels under the maximal error criterion. We see that these codes are optimal by giving a multi-letter characterization of the capacity of compound quantum channels with no assumption on the size of the underlying uncertainty set. We use the asymptotic equivalence of the two tasks of entanglement transmission and entanglement generation to include the capacity region corresponding to simultaneous transmission of classical messages and generation of entanglement between the two parties.

Next, using Ahlswede's robustification and elimination techniques [Ahl78], we convert the codes derived for the compound channel to derive suitable codes for arbitrarily varying quantum channels. This is possible given that the error functions associated with codes corresponding to the compound model decay to zero exponentially. We derive a dichotomy statement [Ahl78] for the simultaneous classical message and entanglement transmission through AVQCs under the average error criterion. This dichotomy is observed when considering two scenarios where the communicating parties do and do not have access to unlimited common randomness, yielding the common-randomness and deterministic capacity regions of the channel model respectively. Therefore, this shows that the common-randomness capacity region of the arbitrarily varying channel is equal to that of the compound channel conv($\mathfrak{J}$), namely the compound channel generated by the convex hull of the uncertainty set of channels $\mathfrak{J}$. Further, if the deterministic capacity of the arbitrarily varying channel is not the point $(0, 0)$, it is equal to the common-randomness capacity of the channel.

We give a necessary and sufficient condition for the deterministic capacity region to be be the point $(0, 0)$. This condition is known as *symmetrizability* of the channel (see [ABBN12] and [BN14]). Finally, we show that the codes derived here can be used for fully quantum AVCs where the jammer is not restricted to product states, but can use general quantum states to parameterize the channel used many times. This model has been introduced in Sect. 4.8.2 along with the main result and related work for fully quantum AVCs, and hence here we avoid further explanation of the techniques used.

The task of simultaneous transmission of classical messages and entanglement was first considered by Devetak and Shor in [DS05] in the case of a memoryless quantum channel under the assumption that the channel state is perfectly known to its users. The authors derived a multi-letter characterization of the capacity region in this setting, which also classified the naïve time-sharing approach as being suboptimal for simultaneous transmission. A code construction sufficient to achieve also the rate pairs lying outside the time-sharing region was derived using a *piggy-backing* technique. A specialized construction introduced in [Dev05] allows us to encode the identity of the classical message into the coding states of an underlying entanglement transmission code. The mentioned strategy, to optimally combine different communication tasks in quantum channel coding, was used and further developed in different directions. We explicitly mention subsequent research activity by Hsieh and Wilde [HW10a, HW10b] where the idea of *piggy backing* classical messages onto quantum codes was extended to include entanglement assistance. The resulting code construction, being sufficient to achieve each point in the three-dimensional rate region for entanglement-assisted classical/quantum simultaneous transmission, leads to a full (multi-letter) characterization of the *quantum dynamic capacity* of a (perfectly known) quantum channel [WH11b] (see the textbook [Wil17] for an up-to-date pedagogical presentation of the mentioned results).

In order to derive classically enhanced quantum codes that are robust against channel uncertainty, we refine the construction entanglement transmission codes for compound quantum channels from [BBN09, BDNW18] instead of elaborating on the usual approach of building upon codes from [Dev05]. In fact, it was noticed earlier that deriving entanglement generation codes from secure classical message transmission codes (the strategy which the arguments in [Dev05] follow) seems not to be suitable when the channel is a compound quantum channel.

In Sect. 4.8.2.1, precise definitions of the channel model, relevant codes used in different scenarios and finally the main results in form of Theorem 4.11 and Theorem 4.12 are given for the compound model. The corresponding definitions and results for the arbitrarily varying model are given in Sect. 4.8.2.4. The entanglement transmission codes used in [BJS19a] are a generalization of the random codes in [BBN09] and [BDNW18], to accommodate conditional typicality of the input on words from many copies of an alphabet. The classical message transmission codes are those from [Mos15] that prove sufficient for our simultaneous coding purposes.

The coding results for the compound model were proved in two steps by authors of [BJS19a]. In the first step, we show that capacity regions that correspond to the case where the sender is restricted to inputting maximally entangled pure states are achieved. In the second step, the authors prove achievability of capacity regions corresponding to general inputs, using elementary methods that are less involved that the usual BSST type results used for this generalization in [BBN09] and [BDNW18].

In [BJS19a], after proving a converse for the capacity region under the arbitrarily varying channel model, the authors prove coding results in this model by converting the compound channel model codes using Ahlswede's *robustification* method. This assumes unlimited common randomness available to the legal parties. They then use an instance of elimination to show that if the deterministic capacity region is not the point $(0, 0)$, a negligible amount of common randomness per use of the channel is sufficient to achieve the same capacity region. They prove further the necessity and sufficiency of the *symmetrizability* condition for the case where the deterministic capacity region is the point $(0, 0)$. Here we only state the results. Finally, in Sect. 4.8.2.5 we generalize these results to the case of a quantum jammer, by stating Theorem 4.13. Here, the jammer is capable of *general attacks*, explained therein.

We consider two channel models of compound and arbitrarily varying quantum channels. They are both generated by an uncertainty set of c.p.t.p. maps. For the purposes of the present section, when considering the arbitrarily varying channel model, we assume finiteness of the generating uncertainty set. This assumption is absent in the case of the compound channel model.

4.8.2.1 The Compound Quantum Channel

Here we consider quantum compound channels. Let $\mathcal{J} := \{\mathcal{N}_s\}_{s \in S} \subset \mathcal{C}(\mathcal{H}_A, \mathcal{H}_B)$ be a set of c.p.t.p. maps. The compound quantum channel generated by $\mathcal{J}$ is given by the family $\{\mathcal{N}^{\otimes n} : \mathcal{N} \in \mathcal{J}\}_{n=1}^{\infty}$. In other words, using n instances of the compound channel is equivalent to using n instances of one of the channels from the uncertainty

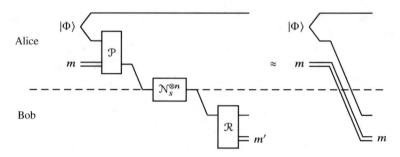

Fig. 4.11 CET coding scheme

set. The users of this channel may or may not have access to the channel state information (CSI). We will often use the set S to index members of $\mathcal{J}$. A compound channel is used $n \in \mathbb{N}$ times by the sender Alice to convey classical messages from a set $[M_{1,n}] := \{1, ..., M_{1,n}\}$ to a receiver Bob. At the same time, the parties would like to communicate quantum information. Here we consider two scenarios in which quantum information can be communicated between the parties.

4.8.2.2   Classically Enhanced Entanglement Transmission (CET)

While transmitting classical messages using $n \in \mathbb{N}$ instances of the compound channel, the sender wishes to transmit the maximally entangled state in her control to the receiver (see Fig. 4.11). The subspace $\mathcal{F}_{A,n}$, with $\mathcal{F}_{A,n} \subset \mathcal{H}_A^{\otimes n}$ and $M_{2,n} := \dim(\mathcal{F}_{A,n})$, quantifies the amount of quantum information transmitted. More precisely:

Definition 4.19   An $(n, M_{1,n}, M_{2,n})$ CET code for $\mathcal{J} \subset \mathcal{C}(\mathcal{H}_A, \mathcal{H}_B)$, is a family

$$\mathcal{C}_{CET} := (\mathcal{P}_m, \mathcal{R}_m)_{m \in [M_{1,n}]} \tag{4.96}$$

with

- $\mathcal{P}_m \in \mathcal{C}(\mathcal{F}_{A,n}, \mathcal{H}_A^{\otimes n})$,
- $\mathcal{R}_m \in \mathcal{C}^{\downarrow}(\mathcal{H}_B^{\otimes n}, \mathcal{F}_{B,n})$ with $\mathcal{F}_{A,n} \subset \mathcal{F}_{B,n}$ and
- $\sum_{m \in [M_{1,n}]} \mathcal{R}_m \in \mathcal{C}(\mathcal{H}_B^{\otimes n}, \mathcal{F}_{B,n})$.

Remark 4.5   We remark that as defined above, for each $m \in [M_{1,n}]$ we have an $(n, M_{2,n})$ entanglement transmission code for $\mathcal{J}$.

For every $m \in M_{1,n}$ and $s \in S$, we define the following performance function for this communication scenario when $n \in \mathbb{N}$ instances of the channel have been used,

$$P(\mathcal{C}_{CET}, \mathcal{N}_s^{\otimes n}, m) := F(|m\rangle \langle m| \otimes \Phi^{AB}, \mathrm{id}_{\mathcal{F}_{A,n}} \otimes \mathcal{R} \circ \mathcal{N}_s^{\otimes n} \circ \mathcal{P}_m(\Phi^{AA})), \tag{4.97}$$

where Φ^{XY} is a maximally entangled state on $\mathcal{F}_{X,n} \otimes \mathcal{F}_{Y,n}$ and

$$\mathcal{R} := \sum_{m \in [M_{1,n}]} |m\rangle \langle m| \otimes \mathcal{R}_m. \tag{4.98}$$

4.8.2.3 Classically Enhanced Entanglement Generation (CEG)

In this scenario, while transmitting classical messages, Alice wishes to establish a pure state shared between her and Bob. As the maximally entangled pure state shared between the parties is an instance of such a pure state, it can be proved that the previous task achieved in CET achieves the task laid out by this one, but the opposite is not necessarily true. More precisely:

Definition 4.20 An $(n, M_{1,n}, M_{2,n})$ CEG code for $\mathcal{J} \subset \mathcal{C}(\mathcal{H}_A, \mathcal{H}_B)$ is a family $\mathcal{C}_{CEG} := (\Psi_m, \mathcal{R}_m)_{m=1}^{M_{1,n}}$, where Ψ_m is a pure state on $\mathcal{F}_{A,n} \otimes \mathcal{H}_A^{\otimes n}$ and

- $\mathcal{R}_m \in \mathcal{C}^{\downarrow}(\mathcal{H}_B^{\otimes n}, \mathcal{F}_{B,n})$ with $\mathcal{F}_{A,n} \subset \mathcal{F}_{B,n}$ and
- $\sum_{m \in [M_{1,n}]} \mathcal{R}_m \in \mathcal{C}(\mathcal{H}_B^{\otimes n}, \mathcal{F}_{B,n})$.

The relevant performance functions for this task, for every $m \in [M_{1,n}]$ and $s \in S$, are

$$P(\mathcal{C}_{CEG}, \mathcal{N}_s^{\otimes n}, m) := F(|m\rangle \langle m| \otimes \Phi^{AB}, \mathrm{id}_{\mathcal{F}_{A,n}} \otimes \mathcal{R} \circ \mathcal{N}_s^{\otimes n}(\Psi_m)), \tag{4.99}$$

with Φ^{AB} maximally entangled on $\mathcal{F}_{A,n} \otimes \mathcal{F}_{B,n}$.

Averaging over the message set $[M_{1,n}]$ will give us the corresponding average performance functions for each $s \in S$,

$$\overline{P}(\mathcal{C}_X, \mathcal{N}_s^{\otimes n}) := \frac{1}{M_{1,n}} \sum_{m \in [M_{1,n}]} P(\mathcal{C}_X, \mathcal{N}_s^{\otimes n}, m) \tag{4.100}$$

for $X \in \{CET, CEG\}$. For each scenario, we define the achievable rates.

Definition 4.21 Let $X \in \{CET, CEG\}$. A pair (R_1, R_2) of non-negative numbers is called an achievable X rate for the compound channel $\mathcal{J}$, if for each $\epsilon, \delta > 0$ exists a number $n_0 = n_0(\epsilon, \delta)$, such that for each $n > n_0$ we find an $(n, M_{1,n}, M_{2,n})$ X code $\mathcal{C}_X$ such that

1. $\frac{1}{n} \log M_{i,n} \geq R_i - \delta$ for $i \in \{1, 2\}$,
2. $\inf_{s \in S} \min_{m \in M_{1,n}} P(\mathcal{C}_X, \mathcal{N}_s^{\otimes n}, m) \geq 1 - \epsilon$

are simultaneously fulfilled. We also define X "average-error-rates" by averaging the performance functions in the last condition over $m \in [M_{1,n}]$. We define the X capacity region of $\mathcal{J}$ by

$$C_X(\mathcal{J}) := \{\,(R_1, R_2) \in \mathbb{R}_0^+ \times \mathbb{R}_0^+ : \qquad\qquad \}. \qquad (4.101)$$

$$(R_1, R_2) \text{ is an achievable X rate for } \mathcal{J}$$

Also the capacity region corresponding to average error criteria is defined as

$$\overline{C}_X(\mathcal{J}) := \{\,(R_1, R_2) \in \mathbb{R}_0^+ \times \mathbb{R}_0^+ : \qquad\qquad \}. \qquad (4.102)$$

$$(R_1, R_2) \text{ is an achievable X average-error-rate for } \mathcal{J}$$

Moreover, let $\mathcal{X}$ be an alphabet, $\mathcal{M} \in \mathcal{C}(\mathcal{H}_A, \mathcal{H}_B)$ $\forall s \in S$, $p \in \mathcal{P}(\mathcal{X})$ and Ψ_x be a pure state for all $x \in \mathcal{X}$. Given the state

$$\omega(\mathcal{M}, p, \Psi) := \sum_{x \in \mathcal{X}} p(x) \, |x\rangle \, \langle x| \otimes \mathrm{id}_{\mathcal{H}_A} \otimes \mathcal{M}(\Psi_x), \qquad (4.103)$$

we introduce the following set,

$$\hat{C}(\mathcal{N}_s, p, \Psi) := \left\{ (R_1, R_2) \in \mathbb{R}_0^+ \times \mathbb{R}_0^+ : \begin{array}{l} R_1 \leq I(X; B, \omega(\mathcal{N}_s, p, \Psi)) \\ \wedge R_2 \leq I(A \rangle BX, \omega(\mathcal{N}_s, p, \Psi)) \end{array} \right\}$$

$$(4.104)$$

with Ψ denoting $(\Psi_x : x \in \mathcal{X})$ collectively. We will also use

$$\frac{1}{l}A := \left\{ \left(\frac{1}{l}x_1, \frac{1}{l}x_2 \right) : (x_1, x_2) \in A \right\}. \qquad (4.105)$$

The following statement is the first main result of this section.

Theorem 4.11 *Let* $\mathcal{J} := \{\mathcal{N}_s\}_{s \in S} \subset \mathcal{C}(\mathcal{H}_A, \mathcal{H}_B)$ *be any compound quantum channel. Then*

$$C_{CET}(\mathcal{J}) = \overline{C}_{CET}(\mathcal{J}) = C_{CEG}(\mathcal{J}) = \overline{C}_{CEG}(\mathcal{J}) \qquad (4.106)$$

$$= \mathrm{cl}\left(\bigcup_{l=1}^{\infty} \frac{1}{l} \bigcup_{p, \Psi} \bigcap_{s \in S} \hat{C}(\mathcal{N}_s^{\otimes l}, p, \Psi) \right) \qquad (4.107)$$

holds.

4.8.2.4   The Arbitrarily-Varying Quantum Channel

The arbitrarily varying quantum channel generated by a set $\mathcal{J} := \{\mathcal{N}_s\}_{s \in S}$ of c.p.t.p. maps with input Hilbert space $\mathcal{H}_A$ and output Hilbert space $\mathcal{H}_B$ is given by the family of c.p.t.p. maps $\{\mathcal{N}_{s^l} : \mathcal{L}(\mathcal{H}_A^{\otimes l}) \to \mathcal{L}(\mathcal{H}_B^{\otimes l}), s^l \in S^l, l \in \mathbb{N}\}_{l=1}^{\infty}$, where

$$\mathcal{N}_{s^l} := \mathcal{N}_{s_1} \otimes \ldots \mathcal{N}_{s_l} \quad (s^l \in S^l). \tag{4.108}$$

We use $\mathcal{J}$ to denote the AVQC generated by $\mathcal{J}$. To avoid further technicalities, we always assume $|S| < \infty$ for the AVQC generating sets appearing in this section. Most of the results in this section may be generalized to the case of general sets by clever use of approximation techniques from convex analysis together with continuity properties of the entropic quantities which appear in the capacity characterizations (see [ABBN12]).

Definition 4.22 An $(l, M_{1,l}, M_{2,l})$ random CET code for $\mathcal{J}$ is a probability measure μ_l on $(\mathcal{C}(\mathcal{F}_{A,l}, \mathcal{H}_A^{\otimes l})^{M_{1,l}} \times \Omega_l, \sigma_l)$, where

- $\Omega_l := \{(\mathcal{R}^{(1)}, \ldots, \mathcal{R}^{(M_{1,l})}), \sum_{m \in [M_{1,l}]} \mathcal{R}^{(m)} \in \mathcal{C}(\mathcal{H}_B^{\otimes l}, \mathcal{F}_{B,l})\}$,
- $\dim(\mathcal{F}_{A,l}) = M_{2,l}, \mathcal{F}_{X,l} \subset \mathcal{H}_X^{\otimes l}, \quad (X \in \{A, B\})$.
- The sigma-algebra σ_l is chosen such that the function

$$g_{s^l}(\mathcal{P}^{(m)}, \mathcal{R}^{(m)}) := F(|m\rangle \langle m| \otimes \Phi^{AB}, \text{id}_{\mathcal{H}_A^{\otimes l}} \otimes \mathcal{R} \circ \mathcal{N}_{s^l} \circ \mathcal{P}^{(m)}(\Phi^{AA})) \tag{4.109}$$

  is measurable with respect to μ_l, for all $m \in [M_{1,l}], s^l \in S^l$. In Eq. (4.109), Φ^{XY} is a maximally entangled state on $\mathcal{F}_{X,l} \otimes \mathcal{F}_{Y,l}$ and $\mathcal{R} := \sum_{m \in [M_{1,l}]} |m\rangle \langle m| \otimes \mathcal{R}^{(m)}$.
- We further require that σ_l contains all the singleton sets. The case where μ_l is deterministic, namely is equal to unity on a singleton set and zero otherwise, gives us a deterministic $(l, M_{1,l}, M_{2,l})$ CET code for $\mathcal{J}$. Abusing the terminology, we also refer to the singleton sets as deterministic codes.

Definition 4.23 A non-negative pair of real numbers (R_1, R_2) is called an achievable CET rate pair for $\mathcal{J} := \{\mathcal{N}_s\}_{s \in S}$ with random codes and average error criterion, if there exists a random CET code μ_l for $\mathcal{J}$ with members of singleton sets notified by $(\mathcal{P}^{(m)}, \mathcal{R}^{(m)})_{m \in [M_{1,l}]}$, such that

$$\liminf_{l \to \infty} \frac{1}{l} \log M_{i,l} \geq R_i \quad (i \in \{1, 2\}), \tag{4.110}$$

$$\lim_{l \to \infty} \inf_{s^l \in S^l} \int \frac{1}{M_{1,l}} \sum_{m \in [M_{1,l}]} g_{s^l}(\mathcal{P}^{(m)}, \mathcal{R}^{(m)}) \, d\mu_l(\mathcal{P}^{(m)}, \mathcal{R}^{(m)})_{m=1}^{M_{1,l}} = 1. \tag{4.111}$$

The random CET capacity region with average error criterion of $\mathcal{J}$ is defined by

$$\overline{\mathcal{A}}_{r,CET}(\mathcal{J}) := \{(R_1, R_2) : (R_1, R_2) \text{ is achievable CET rate pair for } \mathcal{J}$$
$$\text{with random codes and average error criterion}\}.$$

Definition 4.24 A non-negative pair of real numbers (R_1, R_2) is called an achievable deterministic CET rate for $\mathcal{J}$ with average error criterion, if there exists a deterministic $(l, M_{1,l}, M_{2,l})$ CET code $(\mathcal{P}^{(m)}, \mathcal{R}^{(m)})_{m \in [M_{1,l}]}$ for $\mathcal{J}$ with

1. $\liminf_{l \to \infty} \frac{1}{l} \log M_{i,l} \geq R_i$ $(i \in \{1, 2\})$,
2. $\lim_{l \to \infty} \inf_{s^l \in S^l} \frac{1}{M_{1,l}} \sum_{m \in [M_{1,l}]} g_{s^l}(\mathcal{P}^{(m)}, \mathcal{R}^{(m)}) = 1$.

Correspondingly we define the following capacity region,

$$\overline{\mathcal{A}}_{d,CET}(\mathfrak{J}) := \{(R_1, R_2) : (R_1, R_2) \text{ is achievable deterministic}$$

$$CET \text{ rate pair for } \mathfrak{J} \text{ with average error criterion}\}.$$

The deterministic CET codes defined here are entanglement transmission codes for each $m \in [M_{1,l}]$. More precisely, we have the following definition.

Definition 4.25 An (n, M), $n, M \in \mathbb{N}$ entanglement transmission code for AVQC $\mathfrak{J} \subset \mathcal{C}(\mathcal{H}_A, \mathcal{H}_B)$ is a pair $(\mathcal{P}, \mathcal{R})$ with $\mathcal{P} \in \mathcal{C}(\mathcal{F}_{A,n}, \mathcal{H}_A^{\otimes n})$, $\mathcal{R} \in \mathcal{C}(\mathcal{H}_B^{\otimes n}, \mathcal{F}_{B,n})$ with $\mathcal{F}_{A,n} \subset \mathcal{F}_{B,n} \subset \mathcal{H}_A^{\otimes n}$ and $\dim(\mathcal{F}_{A,n}) = M$. The corresponding performance function for this task is

$$F(\Phi^{AB}, \mathrm{id}_{\mathcal{H}_A^{\otimes n}} \otimes \mathcal{R} \circ \mathcal{N}_{s^n} \circ \mathcal{P}(\Phi^{AA})), \quad s^n \in S^n. \tag{4.112}$$

Essential to the statement of our results is the concept of symmetrizablity defined in the following.

Definition 4.26 Let $\mathfrak{J} := \{\mathcal{N}_s\}_{s \in S} \subset \mathcal{C}(\mathcal{H}_A, \mathcal{H}_B)$ with $|S| < \infty$ be an AVQC.

1. $\mathfrak{J}$ is called l-symmetrizable for $l \in \mathbb{N}$, if for each finite set of states $\{\rho_1, \ldots, \rho_K\} \subset \mathcal{S}(\mathcal{H}_A^{\otimes l})$ with $K \in \mathbb{N}$ there is a map $p : \{\rho_1, \ldots, \rho_K\} \to \mathcal{P}(S^l)$ such that for all $i, j \in \{1, \ldots, K\}$

$$\sum_{s^l \in S^l} p(\rho_i)(s^l) \mathcal{N}_{s^l}(\rho_j) = \sum_{s^l \in S^l} p(\rho_j)(s^l) \mathcal{N}_{s^l}(\rho_i). \tag{4.113}$$

2. We call $\mathfrak{J}$ symmetrizable if it is l-symmetrizable for all $l \in \mathbb{N}$.

Remark 4.6 The above definition for *symmetrizability* was first given in [ABBN12], generalizing the concept of symmetrization for classical AVCs from [Eri85]. This definition for *symmetrizability* was meaningfully simplified in [BN14], to require checking of the condition Eq. (4.113) for two input states only (K=2).

The following result is the second main result of this section.

Theorem 4.12 *Let* $\mathfrak{J} := \{\mathcal{N}_s\}_{s \in S} \subset \mathcal{C}(\mathcal{H}_A, \mathcal{H}_B)$ *with* $|S| < \infty$ *be an AVQC. The following hold.*

1. $\overline{\mathcal{A}}_{d,CET}(\mathfrak{J}) \neq \{(0, 0)\}$ *implies*

$$\overline{\mathcal{A}}_{d,CET}(\mathfrak{J}) = \overline{\mathcal{A}}_{r,CET}(\mathfrak{J}) = \overline{C}_{CET}(\mathrm{conv}(\mathfrak{J})), \tag{4.114}$$

where $\overline{C}_{CET}(\mathcal{M})$ *is the CET capacity of compound channel* $\mathcal{M}$ *with average error criterion defined in the previous section and*

$$\text{conv}(\mathcal{J}) := \{\mathcal{N}_q : \mathcal{N}_q := \sum_{s \in S} q(s)\mathcal{N}_s, q \in \mathcal{P}(S)\}. \qquad (4.115)$$

2. $\overline{\mathcal{A}}_{d,CET}(\mathcal{J}) = \{(0,0)\}$ *if and only if $\mathcal{J}$ is symmetrizable.*

4.8.2.5 The Fully-Quantum Arbitrarily-Varying Channel

In this section, we consider simultaneous transmission of classical messages and entanglement over an arbitrarily varying quantum channel with a *quantum jammer*. Let $\mathcal{N} \in \mathcal{C}(\mathcal{H}_A \otimes \mathcal{H}_J, \mathcal{H}_B)$ be a quantum channel whose input space is a tensor product of a Hilbert space $\mathcal{H}_A$ (the legitimate sender's space) and a Hilbert space $\mathcal{H}_J$ which is under control of a quantum jammer. We consider a situation, where for each given block-length n, the jammer may choose any state η on $\mathcal{H}_J^{\otimes n}$ as input in order to disturb the transmission of the legitimate parties.

The fully quantum Arbitrarily Varying Quantum Channel (fully quantum AVQC) generated by $\mathcal{N}$ is given by the family

$$\left\{\mathcal{N}_{n,\sigma}(\cdot) := \mathcal{N}^{\otimes n}(\cdot \otimes \sigma) : \sigma \in \mathcal{S}(\mathcal{H}_J^{\otimes n}), n \in \mathbb{N}\right\} \qquad (4.116)$$

of c.p.t.p. maps. The above channel model has already been under consideration in the case of univariate transmission goals. Karumanchi et al. [KMWY16] utilized the postselection technique from [CKR09] to derive correlated random codes for the fully quantum AVQC from good codes for the compound channel generated by $\mathcal{J} := \{\mathcal{N}_\sigma := \mathcal{N}(\cdot, \sigma) : \sigma \in \mathcal{S}(\mathcal{H}_J)\}$. This approach turned out to be successful to determine the random entanglement transmission capacity for the fully quantum AVQC. In recent work [BDNW18], the above mentioned techniques were also used to characterize the random classical message transmission capacity of the fully quantum AVQC. Going beyond this, the authors of [BDNW18] introduced a derandomization technique to derive a dichotomy for the entanglement and classical message transmission capacities of the fully quantum AVQC. *The deterministic capacity is zero or it equals the random capacity.* We see that the ideas of the mentioned works together with the results presented in this section are sufficient to determine the random capacity and establish a partial characterization of the deterministic capacity in terms of a dichotomy, also in the case of simultaneous transmission of entanglement and classical messages.

The definitions for the corresponding capacity regions can be easily extrapolated from the corresponding definitions in Sect. 4.8.2.4 using the set of transmission maps in Eq. (4.116). We denote the *random CET capacity region* of $\mathcal{N}$ by $\overline{\mathcal{A}}_{r,CET}(\mathcal{N})$ and the *deterministic CET capacity* by $\overline{\mathcal{A}}_{d,CET}(\mathcal{N})$. First, we give a characterization of the random CET capacity $\overline{\mathcal{A}}_{r,CET}(\mathcal{N})$ of the fully quantum AVQC.

Theorem 4.13 *Let* $\mathcal{N} \in \mathcal{C}(\mathcal{H}_A \otimes \mathcal{H}_J, \mathcal{H}_B)$, *and* $\mathfrak{I} := \{\mathcal{N}_\sigma : \sigma \in \mathcal{S}(\mathcal{H}_J)\}$. *It holds*

$$\overline{\mathcal{A}}_{r,CET}(\mathcal{N}) = \overline{\mathcal{C}}_{CET}(\mathfrak{I}) . \tag{4.117}$$

The $\supset$ inclusion in Theorem 4.13 is obvious. To show the reverse inclusion, we will invoke the *robustification* statement in Proposition 4.9 below. In the derivations, the following representation of the permutation group $\mathfrak{S}_n$ on n-fold tensor product spaces plays a key role. Let for each $\pi \in \mathfrak{S}_n$, U_π be the unitary exchanging the factors in $\mathcal{H}^{\otimes n}$, i.e.

$$U_\pi \, x_1 \otimes \cdots \otimes x_n = x_{\pi(1)} \otimes \cdots \otimes x_{\pi(n)} \tag{4.118}$$

for each $x_1, \ldots, x_n \in \mathcal{H}$. We set $\mathcal{U}_\pi(\cdot) := U_\pi(\cdot)U_\pi^*$. In $\mathcal{U}_{A,\pi}$, $\mathcal{U}_{B,\pi}$, $\mathcal{U}_{J,\pi}$ denote the corresponding maps performed on the subsystems under control of A, B, J accordingly. A rather powerful result for states being invariant under permutations of the tensor factors is the following.

Proposition 4.8 (de Finetti Reduction [CKR09]) *Let* $\rho \in \mathcal{S}(\mathcal{H}^{\otimes n})$ *be a permutation invariant, i.e.* $\mathcal{U}_\pi(\rho) = \rho$ *for each* $\pi \in \mathfrak{S}_n$. *It holds*

$$\rho \leq (n+1)^{(\dim \mathcal{H})^2} \int \sigma^{\otimes n} d\mu(\sigma) \tag{4.119}$$

with a probability measure μ.

Proposition 4.9 *Let* $\mathcal{C} := (\mathcal{P}_m, \mathcal{R}_m)_{m=1}^{M_1}$ *be an* (n, M_1, M_2)-*CET code such that with* $\lambda \in (0, 1)$,

$$\inf_{\sigma \in \mathcal{S}(\mathcal{H}_J)} \overline{P}_{CET}(\mathcal{C}, \mathcal{N}_\sigma^{\otimes n})) \geq 1 - \lambda \tag{4.120}$$

holds. With $\mathcal{C}_\pi := (\mathcal{U}_{A,\pi} \circ \mathcal{P}_m, \mathcal{R}_m \circ \mathcal{U}_{B,\pi^{-1}})$ *for each* $\pi \in \mathfrak{S}_n$, *it holds*

$$\inf_{\tau \in \mathcal{S}(\mathcal{H}_J^{\otimes n})} \frac{1}{n!} \sum_{\pi \in \mathfrak{S}_n} \overline{P}_{CET}(\mathcal{C}_\pi, \mathcal{N}_{n,\tau}) \geq 1 - (n+1)^{(\dim \mathcal{H}_J)^2} \cdot \lambda. \tag{4.121}$$

Next we show using a derandomization technique introduced in [BDNW18], the following statement.

Theorem 4.14 (Dichotomy for $\overline{\mathcal{A}}_{d,CET}$**)** $\overline{\mathcal{A}}_{d,CET}(\mathcal{N})$ *equals* $\{(0, 0)\}$ *or* $\overline{\mathcal{A}}_{r,CET}(\mathcal{N})$.

Remark 4.7 The above statement quantifies the deterministic capacity region of the fully quantum AVQC up to a blind spot. It is an open question whether or not there are channels for which $\overline{\mathcal{A}}_{d,CET}(\mathcal{N}) = \{(0, 0)\}$ and $\{(0, 0)\} \subsetneq \overline{\mathcal{A}}_{r,CET}(\mathcal{N})$ does happen.

4.8.3 Compound Quantum Broadcast Channel with Confidential Messages

In this section we consider the compound quantum broadcast channel, connecting one sender to two receivers of different permissions or priorities. The channel is used to perform an integrated task in which a confidential message, kept secret from the third party, is communicated simultaneously with a broadcast message available to both receivers. The requirements on the broadcast message determine two communication scenarios. In the first scenario, we consider the case where both receivers are required to decode the broadcast message. We refer to this message as the common message. In the second scenario the decoding condition is relaxed on one of the receivers. That is, the third party, namely the receiver from whom the confidential message is kept secret, may or may not decode the broadcast message, to which, in this scenario, we refer to as the public message.

The capacity of the channel for performing such tasks will include trade-off regions, determining the resourcefulness of the public/common message transmission capacity, for enhancement of confidential message transmission. Information theoretic analysis of these tasks will naturally be significant when regions beyond those achieved by simple time-sharing between the two tasks are achieved. We first consider the case where the sender is restricted to classical inputs, namely the classical-quantum-quantum (cqq) broadcast model. This model proves useful for obtaining capacity results for the fully quantum broadcast model, where this restriction is lifted.

The classical counterparts of these results are given in [SB14b]. Therein, the authors first derive robust codes for the bidirectional channel, in which both receivers are meant to decode the message. This common message will then piggyback a public message decoded by Bob. The *privacy amplification* strategies are then applied on part of the public codes to obtain information theoretic security via equivocation. In this section, we follow a similar approach in the context of quantum information theory. In [BJS19b], the authors obtain codes for the bidirectional channel (broadcast channel with no security requirement) by generalizing the random codes from [Mos15]. Generalization of these results yields a universal superposition coding for cq channels. The input structure allows us to use privacy amplification arguments ([BCCD14a]) on part of the codebook to achieve the desired privacy rates.

The quantum broadcast model in which the channel is assumed to be perfectly known by communicating parties is considered in [HW09],[WH11a], with and without a pre-shared secret key respectively. Therein, the authors have established a dynamic capacity trade off region using a coding strategy that is channel-dependent. We use a different strategy in which a universal superposition codes for the compound bidirectional channel is established, exploiting properties of Renyi entropies (see [BJS19b]).

Another regime in which the quantum broadcast model with confidential messages has been studied, is the one-shot (single serving) model. A one-shot dynamic capacity theorem was derived for regions corresponding to tasks of common, public and private message transmission over the quantum channel in [SAH$^+$20]. It would be interesting to see if the coding strategies used therein, derived from position based decoding (see [ADJ17, AJW19]), can be used to design codes for the compound channel model.

Precise definitions of channel models, codes and rate regions along with relevant capacity results for the cqq model are given in Sect. 4.8.3.1. The security criterion imposed on the confidential message is the mutual information between Alice and Eve to be arbitrarily small for large numbers of channel uses. As the common, or indeed the public messages are available to Eve, it is required that the mentioned mutual information be conditioned on the broadcast message. Proving the existence of capacity achieving codes was done in [BJS19b], in two steps. First the authors considered the case where there is no security criterion placed on the messages sent to Bob and Eve. In this case, we have a bidirectional channel, where Alice, is sending a message to be decoded by Bob and potentially by Eve (whether Eve decodes this message depends on which scenario is considered, determining, in turn, our labeling of it as common or public). Conditioned on this message (the corresponding codewords are distributed according to a certain structure), Alice is simultaneously transmitting a second type of message that is decoded by Bob. In the second step, the second type of message described above is used for privacy amplification. Finally, we give the code definitions and capacity results for the fully quantum broadcast channel in Sect. 4.8.3.2.

4.8.3.1 Basic Definitions and Main Results

In this section we state the main results and definitions for the compound classical-quantum-quantum (cqq) broadcast channel. The results and definitions related to the fully quantum broadcast channel are stated in Sect. 4.8.3.2. For finite alphabet $\mathcal{X}$ and Hilbert spaces $\mathcal{H}_B, \mathcal{H}_E$, let $\mathcal{W} := \{W_s\}_{s \in S} \subset CQ(\mathcal{X}, \mathcal{H}_B \otimes \mathcal{H}_E)$ be a set of cqq channels. The compound cqq broadcast channel generated by this set is given by the family $\{W_s^{\otimes n}, s \in S\}_{n=1}^{\infty}$. In other words, using n instances of the compound channel is equivalent to using n instances of one of the channels from the uncertainty set. The users of this channel may or may not have access to the channel state information (CSI). Here we consider the case where both users only know the uncertainty set to which the actual channel belongs. We consider two closely related communication scenarios of significance, having both appeared in the literature hitherto.

- **Broadcasting Common and Confidential messages (BCC):** the compound channel is used $n \in \mathbb{N}$ times by the sender Alice in control of the input of the channel, to send two types of messages (m_0, m_c) simultaneously over the channel.

  - $m_0 \in [M_{0,n}]$, called the common message, that has to be reliably decoded by receiver Bob in control of Hilbert space $\mathcal{H}_B$ and Eve in control of Hilbert space $\mathcal{H}_E$.

- $m_c \in [M_{c,n}]$, called the confidential message, that has to be decoded reliably by Bob, while Eve, the wiretapper, is kept ignorant.

- **Transmitting Public and Confidential messages (TPC)**: along with the confidential message $m_c \in [M_{0,n}]$ and instead of the common message, Alice wishes to send a "public" message $m_1 \in [M_{1,n}]$, that is reliably decoded by Bob, while it may or may not be decoded by Eve.

We consider the main concepts and results related to each task in the following. We start with the BCC scenario. The precise definition of the BCC codes is given by the following.

Definition 4.27 (BCC Codes) An $(n, M_{0,n}, M_{c,n})$ BCC code for $\mathcal{W}$ is a family $\mathcal{C} = (E(\cdot|\mathbf{m}), D_{B,\mathbf{m}}, D_{E,m_0})_{\mathbf{m} \in \mathbf{M}}$ with $\mathbf{M} := [M_{0,n}] \times [M_{c,n}]$, stochastic encoder $E : \mathbf{M} \to \mathcal{P}(\mathcal{X}^n)$, POVMs $(D_{B,\mathbf{m}})_{\mathbf{m} \in \mathbf{M}}$ on $\mathcal{H}_B^{\otimes n}$ and $(D_{E,m_0})_{m_0 \in [M_{0,n}]}$ on $\mathcal{H}_E^{\otimes n}$.

We define the transmission error functions for any cqq broadcast channel $W : \mathcal{X} \to \mathcal{S}(\mathcal{H}_B \otimes \mathcal{H}_E)$ and $n \in \mathbb{N}$ by

- $\bar{e}_B(\mathcal{C}, W^{\otimes n}) := \frac{1}{|\mathbf{M}|} \sum_{\mathbf{m} \in \mathbf{M}} \sum_{\mathbf{x} \in \mathcal{X}^n} E(\mathbf{x}|\mathbf{m}) \operatorname{Tr}\left[D_{B,\mathbf{m}}^c W_B^{\otimes n}(\mathbf{x}) \right]$ and
- $\bar{e}_E(\mathcal{C}, W^{\otimes n}) := \frac{1}{|\mathbf{M}|} \sum_{\mathbf{m} \in \mathbf{M}} \sum_{\mathbf{x} \in \mathcal{X}^n} E(\mathbf{x}|\mathbf{m}) \operatorname{Tr}\left[D_{E,m_0}^c W_E^{\otimes n}(\mathbf{x}) \right]$,

where W_γ, $\gamma \in \{B, E\}$ are the marginal channels of W. Moreover, we use the security criterion given by

$$I(M_c; E|M_0, \sigma_{s,n}), \tag{4.122}$$

where $\sigma_{s,n}$ is the code state defined by

$$\sigma_{s,n} := \frac{1}{|\mathbf{M}|} \sum_{\mathbf{m} \in \mathbf{M}} |\mathbf{m}\rangle \langle \mathbf{m}| \otimes \sum_{\mathbf{x} \in \mathcal{X}^n} E(\mathbf{x}|m) W_s^{\otimes n}(\mathbf{x}), \qquad (s \in S, n \in \mathbb{N}). \tag{4.123}$$

The conditional mutual information should be understood given Eq. (4.57), and considering ONBs, $\{|m_i\rangle\}_{m_i \in [M_i]} \in \mathbb{C}^{M_i}$ for $i \in \{0, c\}$ and $|\mathbf{m}\rangle := |m_0\rangle \otimes |m_c\rangle$. Based on this, we define the following achievable rate pairs.

Definition 4.28 (Achievable BCC Rate Pair) A pair (R_0, R_c) of non-negative numbers is called an achievable BCC rate pair for $\mathcal{W}$, if for each $\epsilon, \delta > 0$, exists an $n_0(\epsilon, \delta) \in \mathbb{N}$, such that for all $n > n_0$, we find an $(n, M_{0,n}, M_{c,n})$ BCC code $\mathcal{C} = (E(\cdot|\mathbf{m}), D_{B,\mathbf{m}}, D_{E,m_0})_{\mathbf{m} \in \mathbf{M}}$ such that

1. $\frac{1}{n} \log M_{i,n} \geq R_i - \delta$ $(i \in \{0, c\})$,
2. $\sup_{s \in S} \bar{e}_\gamma(\mathcal{C}, W_s^{\otimes n}) \leq \epsilon$ $(\gamma \in \{B, E\})$,
3. $\sup_{s \in S} I(M_c; E|M_0, \sigma_{s,n}) \leq \epsilon$,

are simultaneously fulfilled.

We define the BCC capacity region of W by

$$C_{BCC}[W] := \{(R_0, R_c) \in \mathbb{R}_0^+ \times \mathbb{R}_0^+ :$$

$$(R_0, R_c) \text{ is an achievable BCC rate pair for } W\}. \tag{4.124}$$

To state our theorem, we define the following regions, given finite alphabets $\mathcal{U}$, $\mathcal{Y}$ and probability distribution $p = p_{UYX} \in \mathcal{P}(\mathcal{U} \times \mathcal{Y} \times \mathcal{X}^n)$, with the random variables U, Y, X distributed accordingly, so that

$$\hat{C}^{(1)}(W, p, n) := \{(R_0, R_c) \in \mathbb{R}_0^+ \times \mathbb{R}_0^+ :$$

$$R_0 \leq \inf_{s \in S} \min \{I(U; B, \omega_s), I(U; E, \omega_s)\} \wedge$$

$$R_c \leq \inf_{s \in S} I(Y; B|U, \omega_s) - \sup_{s \in S} I(Y; E|U, \omega_s)\}, \tag{4.125}$$

with

$$\omega_s := \sum_{(u,y,\mathbf{x}) \in \mathcal{U} \times \mathcal{Y} \times \mathcal{X}^n} p(u, y, \mathbf{x}) |u\rangle \langle u| \otimes |y\rangle \langle y| \otimes W_s^{\otimes n}(\mathbf{x}). \tag{4.126}$$

We state the following theorem.

Theorem 4.15 Let $W := \{W_s\}_{s \in S} \subset CQ(\mathcal{X}, \mathcal{H}_B \otimes \mathcal{H}_E)$ be any compound cqq broadcast channel. It holds

$$C_{BCC}[W] = \text{cl}\left(\bigcup_{l=1}^{\infty} \bigcup_p \frac{1}{l} \hat{C}^{(1)}(W, p, l)\right), \tag{4.127}$$

where we have used $\frac{1}{l} A := \{(\frac{1}{l}x_1, \frac{1}{l}x_2) : (x_1, x_2) \in A\}$. The second union is taken over all $p_{UYX} \in \mathcal{P}(\mathcal{U} \times \mathcal{Y} \times \mathcal{X}^l)$ such that random variable $U - Y - X$ form a Markov chain and alphabets $\mathcal{U}$ and $\mathcal{Y}$ are finite.

Remark 4.8 The set given on the right hand side of Eq. (4.127) is convex and hence we do not need further *convexification* here. This results from time sharing arguments applied on the entropic quantities appearing in Eq. (4.127). For a short proof of a similar statement, see [BJS19a].

We proceed with the TPC scenario. The precise definition of the TPC codes is given in the following.

Definition 4.29 (TPC Codes) An $(n, M_{1,n}, M_{c,n})$ TPC code for W is a family $\mathcal{C} = (E(\cdot|\mathbf{m}), D_{B,\mathbf{m}})_{\mathbf{m} \in \mathbf{M}}$ with $\mathbf{M} := [M_{1,n}] \times [M_{c,n}]$, stochastic encoder $E : \mathbf{M} \to \mathcal{P}(\mathcal{X}^n)$ and a POVM $(D_{B,\mathbf{m}})_{\mathbf{m} \in \mathbf{M}}$ on $\mathcal{H}_B^{\otimes n}$.

We define the relevant transmission error function for any cqq broadcast channel $W : \mathcal{X} \to \mathcal{S}(\mathcal{H}_B \otimes \mathcal{H}_E)$ and $n \in \mathbb{N}$ by

$$\bar{e}_B(\mathcal{C}, W^{\otimes n}) := \frac{1}{|\mathbf{M}|} \sum_{\mathbf{m} \in \mathbf{M}} \sum_{\mathbf{x} \in \mathcal{X}^n} E(\mathbf{x}|\mathbf{m}) \operatorname{Tr}\left[D^c_{B,\mathbf{m}} W_B^{\otimes n}(\mathbf{x})\right]. \tag{4.128}$$

Moreover, we use the security criterion given by

$$I(M_c; E|M_1, \sigma_{s,n}), \tag{4.129}$$

where $\sigma_{s,n}$ is the code state defined by

$$\sigma_{s,n} := \frac{1}{|\mathbf{M}|} \sum_{\mathbf{m} \in \mathbf{M}} |\mathbf{m}\rangle \langle \mathbf{m}| \otimes \sum_{\mathbf{x} \in \mathcal{X}^n} E(\mathbf{x}|\mathbf{m}) W_s^{\otimes n}(\mathbf{x}). \tag{4.130}$$

Again, we note that the conditional mutual information should be understood given Eq. (4.57) and considering ONBs $\{|m_i\rangle\}_{m_i \in [M_i]} \in \mathbb{C}^{M_i}$ for $i \in \{1, c\}$ and $|\mathbf{m}\rangle := |m_1\rangle \otimes |m_c\rangle$. Based on this, we define the following achievable rate pairs.

Definition 4.30 (Achievable TPC Rate Pair) A pair (R_1, R_c) of non-negative numbers is called an achievable TPC rate pair for $\mathcal{W}$, if for each $\epsilon, \delta > 0$, exists an $n_0(\epsilon, \delta) \in \mathbb{N}$, such that for all $n > n_0$ we find an $(n, M_{1,n}, M_{c,n})$ TPC code $\mathcal{C} = (E(\cdot|\mathbf{m}), D_{B,\mathbf{m}})_{\mathbf{m} \in \mathbf{M}}$, such that

1. $\frac{1}{n} \log M_{i,n} \geq R_i - \delta$ ($i \in \{1, c\}$),
2. $\sup_{s \in S} \bar{e}_B(\mathcal{C}, W_s^{\otimes n}) \leq \epsilon$,
3. $\sup_{s \in S} I(M_c; E|M_1, \sigma_{s,n}) \leq \epsilon$

are simultaneously fulfilled.

We define the TPC capacity region of $\mathcal{W}$ by

$$C_{TPC}[\mathcal{W}] := \{(R_1, R_c) \in \mathbb{R}_0^+ \times \mathbb{R}_0^+ :$$
$$(R_1, R_c) \text{ is an achievable TPC rate for } \mathcal{W}\}. \tag{4.131}$$

To state our theorem, we define the following sub-regions, given finite alphabets $\mathcal{V}, \mathcal{Y}$ and probability distribution $p = p_{VYX} \in \mathcal{P}(\mathcal{V} \times \mathcal{Y} \times \mathcal{X}^n)$, with the random variables V, Y, X distributed accordingly, then

$$C^{(1)}(\mathcal{W}, p, n) := \{(R_1, R_c) \in \mathbb{R}_0^+ \times \mathbb{R}_0^+ : R_1 \leq \inf_{s \in S} I(V; B, \omega_s) \wedge$$
$$R_c \leq \inf_{s \in S} I(Y; B|V, \omega_s) - \sup_{s \in S} I(Y; E|V, \omega_s)\}, \tag{4.132}$$

with

$$\omega_s := \sum_{(v,y,\mathbf{x}) \in \mathcal{V} \times \mathcal{X} \times \mathcal{X}} p(v, y, \mathbf{x}) |v\rangle \langle v| \otimes |y\rangle \langle y| \otimes W_s^{\otimes n}(\mathbf{x}). \tag{4.133}$$

We can state the following theorem.

Theorem 4.16 *Let* $\mathcal{W} := \{W_s\}_{s \in S} \subset CQ(\mathcal{X}, \mathcal{H}_B \otimes \mathcal{H}_E)$ *be any compound cqq broadcast channel. It holds*

$$C_{TPC}[\mathcal{W}] = cl\left(\bigcup_{l=1}^{\infty}\bigcup_{p}\frac{1}{l}C^{(1)}(\mathcal{W}, p, l)\right). \tag{4.134}$$

The second union is taken over all $p_{VYX} \in \mathcal{P}(\mathcal{V} \times \mathcal{Y} \times \mathcal{X}^l)$ *such that random variable* $V - Y - X$ *form a Markov chain and alphabets* $\mathcal{V}$ *and* $\mathcal{Y}$ *are finite.*

Again, we note Remark 4.8, regarding convexity of the set on the right hand side of Eq. (4.134).

4.8.3.2 BCC and TPC Capacities of Compound Quantum Broadcast Channels

In this section we state results from [BJS19b], on the *fully quantum* setting where the receivers input quantum systems to the channels, i.e. the transition maps of the channels are c.p.t.p. maps instead of cq channels. Since the message transmission tasks we aim to perform are of a classical nature, the corresponding coding theorems can be proved applying the results from earlier chapters.

Explicitly, we apply the results of the preceding sections to derive codes for full quantum broadcast channels. For the remainder of this section, we fix an arbitrary set $\mathcal{J} := \{\mathcal{N}_s\}_{s \in S}$, where

$$\mathcal{N}_s : \mathcal{L}(\mathcal{H}_A) \rightarrow \mathcal{L}(\mathcal{H}_B \otimes \mathcal{H}_E) \tag{4.135}$$

is a c.p.t.p. map for each $s \in S$. Traditionally, the c.p.t.p. map $\mathcal{N}_s$ is assumed to be an isometric channel, namely a Stinespring isometry to a given channel connecting A and B. This way of defining the channel is fairly justified, since it naturally equips E with the strongest abilities when attacking the confidential transmission goals of the remaining parties. However, dropping this assumption on the channel does not complicate any subsequent arguments.

In what follows, we consider the BCC scenario. Corresponding considerations regarding the TPC scenario are easily extrapolated and are hence left to the reader.

Definition 4.31 (BCC Codes) An (n, M_0, M_c) *BCC code* for $\mathcal{J}$ for channels in $\mathcal{C}(\mathcal{H}_A, \mathcal{H}_B \otimes \mathcal{H}_E)$ is a family $\mathcal{C} = (V(m), D_{B,m}, D_{E,m_0})_{m \in \mathbf{M}}$ with $\mathbf{M} := [M_0] \times [M_c]$, where $(D_{B,m})_{m \in \mathbf{M}}$ and $(D_{E,m_0})_{m_0 \in [M_0]}$ are POVMs on $\mathcal{H}_B^{\otimes n}$ and $\mathcal{H}_E^{\otimes n}$ respectively and $V(m)$ is a state on $\mathcal{H}_A^{\otimes n}$ for each m.

The average transmission errors for the receivers B, and E with channel $\mathcal{N} : \mathcal{L}(\mathcal{H}_A) \rightarrow \mathcal{L}(\mathcal{H}_B \otimes \mathcal{H}_E)$ and (n, M_0, M_c)-code $\mathcal{C}$ are defined by

$$\bar{e}_B(\mathcal{C}, \mathcal{N}^{\otimes n}) := \frac{1}{|\mathbf{M}|} \sum_{m \in \mathbf{M}} \text{Tr}\left[D^c_{B,m} \mathcal{N}^{\otimes n}(V(m))\right], \tag{4.136}$$

and

$$\bar{e}_E(\mathcal{C}, \mathcal{N}^{\otimes n}) := \frac{1}{|\mathbf{M}|} \sum_{m \in \mathbf{M}} \text{Tr}\left[D^c_{E,m_0} \mathcal{N}^{\otimes n}(V(m))\right]. \tag{4.137}$$

By replacing the code and errors, the definitions of achievable rate pairs can be directly guessed from Definition 4.28 (the notational ambiguity should cause no misunderstanding since the set $\mathcal{J}$ determines whether the classical-quantum or quantum broadcast channel scenario are considered). We denote the corresponding BCC capacity region by $C_{BCC}[\mathcal{J}]$. Moreover, we define $\hat{C}^{(1)}(\mathcal{J}, p, l, (\rho_y)_{y \in \mathcal{Y}})$, the set of all points in $\mathbb{R}^2$ which fulfil the inequalities

$$0 \leq R_0 \leq \inf_{s \in S} \min \{I(U; B, \omega_s), I(U; E, \omega_s)\} \tag{4.138}$$

and

$$0 \leq R_c \leq \inf_{s \in S} I(Y; B|U, \omega_s) - \inf_{s \in S} I(Y; E|U, \omega_s), \tag{4.139}$$

where we understand the entropic quantities above as being evaluated on the ccq state

$$\omega_s := \omega(\mathcal{N}_s, p, l) := \sum_{u \in \mathcal{U}, y \in \mathcal{Y}} P_{UY}(u, y) \cdot |u, y\rangle \langle u, y| \otimes \mathcal{N}^{\otimes l}_s(\rho_y) \tag{4.140}$$

for each $s \in S$.

Theorem 4.17 *It holds*

$$C_{BCC}[\mathcal{J}] = \text{cl}\left(\bigcup_{l=1}^{\infty} \bigcup_p \frac{1}{l} \hat{C}^{(1)}(\mathcal{J}, p, l)\right) \tag{4.141}$$

The second union is taken over all $p_{UYX} \in \mathcal{P}(\mathcal{U} \times \mathcal{Y} \times \mathcal{X}^l)$, such that the random variable $U - Y - X$ distributed accordingly form a Markov chain and alphabets $\mathcal{U}$ and $\mathcal{Y}$ are finite.

4.8.4 Robust Secure Message Transmission over the Wiretap Channel with a Jammer

In this section we present the results for the tasks introduced in Sect. 4.6, to quantify the capacities achievable by codes that are robust to channel uncertainties considered in the compound and arbitrarily varying models.

The results for transmission capacities can be found in [BCCD14a, BCDN16]. In [BCCD14a] the security of the classical compound channel with quantum wiretapper and channel state information (CSI) at the transmitter was derived. Furthermore, a lower bound on the secure capacity of this channel without CSI and the secure capacity of the compound classical-quantum wiretap channel with CSI at the transmitter is determined. In [Mos15], a multi-letter formula for the secure capacity of the compound classical-quantum wiretap channel is given. The authors of [BDW19] showed that the capacity of a compound wiretap cqq-channel again satisfies a dichotomy theorem.

Definition 4.32 Let Θ and Σ be index sets and let $\mathcal{W} = \{W_t : \mathcal{X} \to \mathcal{S}(\mathcal{H}_B) : t \in \Theta\}$ and $\mathcal{V} = \{V_s : \mathcal{X} \to \mathcal{S}(\mathcal{E}) : s \in \Sigma\}$ be compound cq-channels. We call the pair $(\mathcal{W}, \mathcal{V})$ a *compound wiretap cqq-channel*. The channel output of $\mathcal{W}$ is available to the legitimate receiver (Bob) and the channel output of $\mathcal{V}$ is available to the wiretapper (Eve). We sometimes write the channel as a family of pairs $(\mathcal{W}, \mathcal{V}) = (W_t, V_s)_{t\in\Theta, s\in\Sigma}$.

Definition 4.33 An (n, M, λ) transmission code for the compound wiretap cqq-channel $(W_t, V_s)_{t\in\Theta, s\in\Sigma}$ consists of a family $\mathcal{C} = (P_i, D_i)_{i\in[M]}$, where the P_i are probability distributions on $\mathcal{X}^n$ and $(D_i)_{i\in[M]}$ a POVM on $\mathcal{H}_B{}^{\otimes n}$, such that

$$\forall i \in [M] \quad \sup_{t\in\Theta} 1 - \text{Tr}\left[W^{\otimes n}(P_i)\cdot D_i\right] \le \lambda, \tag{4.142}$$

$$\forall i, j \in [M] \quad \sup_{s\in\Sigma} \frac{1}{2}\|V_s^{\otimes n}(P_i) - V_s^{\otimes n}(P_j)\|_1 \le \mu. \tag{4.143}$$

The capacity is defined as before.

Theorem 4.18 ([BCDN16]) *The secure capacity of a compound wiretap cqq-channel $(\mathcal{W}, \mathcal{V})$ is given by*

$$C_S(\mathcal{W}, \mathcal{V}) = \lim_{n\to\infty} \sup_{U\to X^n\to(B_t^n E_s^n)}$$

$$\frac{1}{n}\left(\inf_{t\in\Theta} I(U; B_t^n) - \sup_{s\in\Sigma} I(U; E_s^n)\right), \tag{4.144}$$

where B_t are the resulting random quantum states at the output of legal receiver channels and E_s are the resulting random quantum states at the output of the wiretap channel.

We turn to the arbitrarily varying model, where along with wiretapper, we also have a jammer. We start this by considering the wiretap model for public message transmission.

Definition 4.34 Let Θ be a finite index set, $\mathcal{X}$ a finite set and $\mathcal{H}_B$ a finite-dimensional Hilbert space. Let $W_t : \mathcal{X} \longrightarrow \mathcal{S}(\mathcal{H}_B)$ be a cq-channel for every $t \in \Theta$:

$$W_t : \mathcal{X} \ni x \mapsto W_t(x) \in \mathcal{S}(\mathcal{H}_B), \quad t \in \Theta. \tag{4.145}$$

Let $t^n \in \Theta^n$ be a state sequence. The memoryless extension of the cq-channel W_{t^n} is given by $W_{t^n}(x^n) = W_{t_1}(x_1) \otimes \ldots \otimes W_{t_n}(x_n)$ for $x^n \in \mathcal{X}^n$. We call $\mathcal{W} = \{W_t\}_{t \in \Theta}$ *an arbitrarily varying cq-channel.*

In this case a jammer can change the channel during the transmission.

Definition 4.35 An (n, M, λ)-*code* for the arbitrarily varying cq-channel $\mathcal{W}$ is a family $\mathcal{C} = ((P_m, D_m) : m \in [M])$ consisting of pairs of stochastic encodings given by code word probability distributions P_m over $\mathcal{X}^n$ and positive semi-definite operators D_i on $\mathcal{H}_B^{\otimes n}$, forming a POVM, i.e. $\sum_{m=1}^{M} D_m = \mathbb{1}$, such that

$$\max_{t^n \in \Theta^n} \max_{i \in [M]} 1 - \mathrm{Tr}\left[W_{t^n}^{\otimes n}(P_i)D_i\right] \leq \lambda. \tag{4.146}$$

Like in the compound case, here we allow explicitly stochastic encoders. The number M is called the *size* of the code, and λ the error probability. The maximum M for given n and λ is denoted $M(n, \lambda)$, extending the definition for a cq-channel (which is recovered for $|\Theta| = 1$).

The capacity of $\mathcal{W}$ is defined as before,

$$C(\mathcal{W}) = \inf_{\lambda > 0} \liminf_{n \to \infty} \frac{1}{n} \log M(n, \lambda). \tag{4.147}$$

A more intuitive description of the arbitrarily varying cq-channel is that a jammer tries to prevent the legal parties from communicating properly. He may change his input in every channel use and is not restricted to use a repetitive probabilistic strategy. Quite on the contrary, it is understood that the sender and the receiver have to select their coding scheme first. After that the jammer makes his choice of the sequence of channel states. The sender and receiver do not know which channel from the set $\mathcal{W}$ is actually used; their prior knowledge is merely that the channel is memoryless and belongs to the set $\mathcal{W}$. Their task is to prepare for the worst case among those. To state dichotomy results, again we need the concept of symmetrizability (see Definition 4.26).

Definition 4.36 We say that the arbitrarily varying cq-channel $\mathcal{W} = \{W_t : t \in \Theta\}$ is *symmetrizable* if there exists a parametrized set of distributions $\{\tau(\cdot|x) : x \in \mathcal{X}\}$ on Θ, also known as a channel τ from $\mathcal{X}$ to Θ, such that for all $x, x' \in \mathcal{X}$,

$$\sum_{t \in \Theta} \tau(t|x) W_t(x') = \sum_{t \in \Theta} \tau(t|x') W_t(x). \tag{4.148}$$

To formulate the capacity theorem of [AB07], we need the following notations. For an arbitrarily varying cq-channel W we denote its convex hull by $\mathrm{conv}(W)$. It is defined as follows:

$$\mathrm{conv}(W) = \left\{ W_q : W_q = \sum_{t \in \Theta} q(t) W_t, \ q \in \mathcal{P}(\Theta) \right\}. \tag{4.149}$$

Furthermore, we set

$$C_{\mathrm{ran}}(W) = \max_{p \in \mathcal{P}(\mathcal{X})} \min_{W \in \mathrm{conv}(W)} I(p; W). \tag{4.150}$$

This is called the random coding capacity of the channel. Under this notion, the encoding with a stochastic encoder is generalized to a (correlated) random code. It is assumed that the sender and the receiver have access to some source with correlated randomness, which, however, is secret from the jammer. Here we need just the quantity to give the transmission capacity of the arbitrarily varying cq-channel.

Theorem 4.19 ([AB07]) *Let W be an arbitrarily varying cq-channel. Then its capacity $C(W)$ is given by*

$$C(W) = \begin{cases} 0 & \text{if } W \text{ is symmetrizable,} \\ C_{ran}(W) & \text{otherwise.} \end{cases} \quad \blacksquare \tag{4.151}$$

At this point we can introduce the wiretapper to the picture. First we define the transmission codes and quote the known transmission capacity.

Definition 4.37 Let Θ and Σ be finite index sets, and let $W = \{W_t : \mathcal{X} \to \mathcal{S}(\mathcal{H}_B) : t \in \Theta\}$ and $V = \{V_s : \mathcal{X} \to \mathcal{S}(\mathcal{E}) : s \in \Sigma\}$ be arbitrarily varying cq-channels. We call the pair (W, V) an *arbitrarily varying wiretap cqq-channel*. The channel output of W is available to the legitimate receiver (Bob) and the channel output of V is available to the wiretapper (Eve). We may sometimes write the channel as a family of pairs $(W, V) = (W_t, V_s)_{t \in \Theta, s \in \Sigma}$.

Definition 4.38 An (n, M, λ) transmission code for the arbitrarily varying wiretap cqq-channel $(W_t, V_s)_{t \in \Theta, s \in \Sigma}$ consists of a family $\mathcal{C} = (P_i, D_i)_{i \in [M]}$, where the P_i are probability distributions on $\mathcal{X}^n$ and $(D_i)_{i \in [M]}$ a POVM on $\mathcal{H}_B^{\otimes n}$ such that

$$\forall i \in [M] \quad \max_{t^n \in \Theta^n} 1 - \mathrm{Tr} \, W_{t^n}^{\otimes n}(P_i) \cdot D_i \leq \lambda, \tag{4.152}$$

$$\forall i, j \in [M] \quad \max_{s^n \in \Sigma^n} \frac{1}{2} \| V_{s^n}^{\otimes n}(P_i) - V_{s^n}^{\otimes n}(P_j) \|_1 \leq \mu. \tag{4.153}$$

The capacity is defined as before. To state the result of [BCND19], we again introduce the random coding capacity,

$$C_{S,\text{ran}}(\mathcal{W}, \mathcal{V}) = \lim_{n \to \infty} \frac{1}{n} \max_{U \to X^n \to B_{t^n}^n E_{s^n}^n}$$

$$\left(\min_{W \in \text{conv}\{W_t : t \in \Theta\}} I(p_U; W^{\otimes n}) - \max_{s^n \in \Sigma^n} I(p_U; V_{s^n}) \right). \qquad (4.154)$$

Here, $B_{t^n}^n$ are the resulting quantum states at the output of the legitimate receiver's channels. $E_{s^n}^n$ are the resulting quantum states at the output of the wiretap channels. The maximum is taken over all random variables that satisfy the Markov chain relationships: $U \to X^n \to B_{t^n}^n E_{s^n}^n$. X^n is here a random variable taking values in $\mathcal{X}^n$, U a random variable taking values on some finite set $\mathcal{U}$ with probability distribution p_U. In [BCND19] the following dichotomy is shown.

Theorem 4.20 ([BCND19]) *Let $C_S(\mathcal{W}, \mathcal{V})$ denote the capacity of the arbitrarily varying wiretap cqq-channel $(\mathcal{W}, \mathcal{V})$. Then,*

$$C_S(\mathcal{W}, \mathcal{V}) = \begin{cases} 0 & \text{if } W \text{ is symmetrizable,} \\ C_{S,\text{ran}}(\mathcal{W}, \mathcal{V}) & \text{otherwise.} \end{cases} \qquad (4.155) \quad \blacksquare$$

Based on these results, one can determine capacity results for public and secure identification.

4.8.5 Robust Identification over CQ Channel for Public and Secure Communication

In this section we present the results for the tasks introduced in Sect. 4.7, to quantify the capacities achievable by codes that are robust to channel uncertainties considered in the compound and arbitrarily varying models.

4.8.5.1 Identification Over Compound CQ Channel

Here we will define the identification capacity of a compound cq-channel and give its single-letter formula. In [Mos15] and [BJK17], the transmission capacity was derived. The authors of [BDW19] used the transmission codes to build an identification code with the method introduced in [AD89a]. This method was also used in [L99] to get the identification capacity of a cq-channel. For the converse, the authors of [BDW19] generalize the method of [AW02].

Definition 4.39 An $(n, N, \lambda_1, \lambda_2)$ *ID-code* for the compound cq-channel $\mathcal{W}$ is a set of pairs $\{(P_i, D_i) : i \in [N]\}$, where the P_i are probability distributions on $\mathcal{X}^n$ and the D_i are POVM elements, i.e. $0 \leq D_i \leq \mathbb{1}$ acting on $\mathcal{H}_B^{\otimes n}$, such that $\forall i \neq j \in [N]$

$$\inf_{t \in \Theta} \mathrm{Tr}\left[W_t^{\otimes n}(P_i) \cdot D_i \right] \geq 1 - \lambda_1 \tag{4.156}$$

and

$$\sup_{t \in \Theta} \mathrm{Tr}\left[W_t^{\otimes n}(P_i) \cdot D_j \right] \leq \lambda_2. \tag{4.157}$$

The largest size of an $(n, N, \lambda_1, \lambda_2)$ ID-code is denoted $N(n, \lambda_1, \lambda_2)$. Analogous to previous definitions, we also have simultaneous ID-codes and the maximum code size $N_{\mathrm{sim}}(n, \lambda_1, \lambda_2)$.

The identification capacities are defined as before in Sect. 4.7. We have considered the so-called pessimistic definitions of capacity thus far. The optimistic definition of capacity is $\bar{C}(\mathcal{W}) = \inf_{\lambda>0} \limsup_{n\to\infty} \frac{1}{n} \log M(n, \lambda)$. It was shown by authors of [BDW19] that the converse holds for the optimistic definition and therefore for the pessimistic definition also.

Theorem 4.21 *Let $\mathcal{W}$ be an arbitrary compound cq-channel with finite input alphabet $\mathcal{X}$ and finite-dimensional output Hilbert space $\mathcal{H}_B$. Then,*

$$C_{\mathrm{ID}}(\mathcal{W}) = C_{\mathrm{ID}}^{\mathrm{sim}}(\mathcal{W}) = C(\mathcal{W}) = \inf_{t \in \Theta} I(Q; W_t), \tag{4.158}$$

and the weak converse holds for the optimistic ID-capacity. Indeed,

$$\inf_{\lambda_1, \lambda_2 > 0} \limsup_{n \to \infty} \frac{1}{n} \log N(n, \lambda_1, \lambda_2) \tag{4.159}$$

$$= \inf_{\lambda_1, \lambda_2 > 0} \liminf_{n \to \infty} \frac{1}{n} \log N_{\mathrm{sim}}(n, \lambda_1, \lambda_2) \tag{4.160}$$

$$= C(\mathcal{W}). \tag{4.161}$$

4.8.5.2 Identification over the CQQ Wiretap Channel

In this section we consider robust secure identification over cqq wiretap channels. First we define public identification codes.

Definition 4.40 An $(n, N, \lambda_1, \lambda_2, \mu)$ *wiretap ID-code* for the compound wiretap cqq-channel $(\mathcal{W}, \mathcal{V})$ is a set of pairs $\{(P_i, D_i) : i \in [N]\}$ where the P_i are probability distributions on $\mathcal{X}^n$ and the $D_i, 0 \leq D_i \leq \mathbb{1}$ denote operators on $\mathcal{H}_B^{\otimes n}$, such that for all $i \neq j \in [N]$ and all $0 \leq F \leq \mathbb{1}$,

$$\inf_{t \in \Theta} \mathrm{Tr}\left[W_t^{\otimes n}(Q_i)D_i\right] \geq 1 - \lambda_1, \tag{4.162}$$

$$\sup_{t \in \Theta} \mathrm{Tr}\left[W_t^{\otimes n}(Q_j)D_i\right] \leq \lambda_2, \tag{4.163}$$

$$\inf_{s \in \Sigma} \mathrm{Tr}\left[V_s^{\otimes n}(Q_j)F\right] + \mathrm{Tr}\left[V_s^{\otimes n}(Q_i)(\mathbb{1} - F)\right] \geq 1 - \mu. \tag{4.164}$$

We define $N(n, \lambda_1, \lambda_2, \mu)$ as the largest N satisfying the above definition for a given n and set $\lambda_1, \lambda_2, \mu$ of errors.

Definition 4.41 The secure identification capacity $C_{SID}(\mathcal{W}, \mathcal{V})$ of a compound wiretap cqq-channel $(\mathcal{W}, \mathcal{V})$ is defined as

$$C_{SID}(\mathcal{W}, \mathcal{V}) := \inf_{\lambda_1, \lambda_2, \mu > 0} \liminf_{n \to \infty} \frac{1}{n} \log \log N(n, \lambda_1, \lambda_2, \mu). \tag{4.165}$$

Again we get a dichotomy result.

Theorem 4.22 *Let $(\mathcal{W}, \mathcal{V})$ be a compound wiretap cqq-channel. Then,*

$$C_{SID}(\mathcal{W}, \mathcal{V}) = C_{SID}^{\mathrm{sim}}(\mathcal{W}, \mathcal{V}) \tag{4.166}$$

$$= \begin{cases} C(\mathcal{W}) & \text{if } C_S(\mathcal{W}, \mathcal{V}) > 0, \\ 0 & \text{if } C_S(\mathcal{W}, \mathcal{V}) = 0. \end{cases} \tag{4.167}$$

At this point we introduce the jammer to the picture.

4.8.5.3   Identification in the Presence of a Jammer

In this section we perform the same analysis for the case of arbitrarily varying cq-channels with analogous findings.

Definition 4.42 An $(n, N, \lambda_1, \lambda_2)$ *ID-code* for the arbitrarily varying cq-channel $\mathcal{W}$ is a set of pairs $\{(P_i, D_i) : i \in [N]\}$, where the P_i are probability distributions on $\mathcal{X}^n$ and the D_i are POVM elements, i.e. $0 \leq D_i \leq \mathbb{1}$, acting on $\mathcal{H}_B^{\otimes n}$, such that $\forall i \neq j \in [N]$

$$\min_{t^n \in \Theta^n} \mathrm{Tr}\, W_{t^n}(P_i) \cdot D_i \geq 1 - \lambda_1 \text{ and} \tag{4.168}$$

$$\max_{t^n \in \Theta^n} \mathrm{Tr}\, W_{t^n}(P_i) \cdot D_j \leq \lambda_2. \tag{4.169}$$

The largest size of an $(n, N, \lambda_1, \lambda_2)$ ID-code is denoted $N(n, \lambda_1, \lambda_2)$. Analogous to previous definitions, we have also simultaneous ID-codes and the maximum code size $N_{\mathrm{sim}}(n, \lambda_1, \lambda_2)$.

The identification capacities are now defined as before. With the help of the method from proof of Theorem 4.21, the authors of [BDW19] showed that the transmission capacity of the channel corresponds to the identification capacity. To do this, in the proof of the direct part, they used a code for an arbitrary varying cq-channel instead of the code for the compound cq-channel. To show the converse, they showed that the error of the first type in the identification can not be arbitrarily small if the channel is *symmetrizable*, and therefore, obtained the following.

Theorem 4.23 *Let W be an arbitrarily varying cq-channel. Then its ID-capacity is given by*

$$C_{ID}^{sim}(W) = C_{ID}(W)$$

$$= \begin{cases} 0 & \text{if } W \text{ is symmetrizable,} \\ C_{ran}(W) & \text{otherwise.} \end{cases} \qquad \blacksquare \qquad (4.170)$$

4.8.5.4   Secure Identification in the Presence of a Jammer

In this section we add a wiretapper to the arbitrarily varying cq-channel. The ID codes are defined as follows.

Definition 4.43 An $(n, N, \lambda_1, \lambda_2, \mu)$ *wiretap ID-code* for the arbitrarily varying wiretap cqq-channel (W, V) is a set of pairs $\{(P_i, D_i) : i \in [N]\}$ where the P_i are probability distributions on X^n and the D_i, $0 \leq D_i \leq \mathbb{1}$ denote operators on $\mathcal{H}_B^{\otimes n}$ such that $\forall i, j \in [N]$, $i \neq j$ and $0 \leq F \leq \mathbb{1}$,

$$\min_{t^n \in \Theta^n} \text{Tr}\left[W_{t^n}^{\otimes n}(Q_i)D_i\right] \geq 1 - \lambda_1, \qquad (4.171)$$

$$\max_{t^n \in \Theta^n} \text{Tr}\left[W_{t^n}^{\otimes n}(Q_j)D_i\right] \leq \lambda_2, \qquad (4.172)$$

$$\min_{s^n \in \Sigma^n} \text{Tr}\left[V_{s^n}^{\otimes n}(Q_j)F\right] + \text{Tr}\left[V_{s^n}^{\otimes n}(Q_i)(\mathbb{1} - F)\right] \geq 1 - \mu. \qquad (4.173)$$

We define $N(n, \lambda_1, \lambda_2, \mu)$ as the largest N satisfying the above definition for a given n and set $\lambda_1, \lambda_2, \mu$ of errors.

Definition 4.44 The identification capacity $C_{SID}(W, V)$ of an arbitrarily varying wiretap cqq-channel (W, V) is defined as

$$C_{SID}(W, V)$$

$$= \inf_{\lambda_1, \lambda_2, \mu > 0} \liminf_{n \to \infty} \frac{1}{n} \log \log N(n, \lambda_1, \lambda_2, \mu). \qquad (4.174)$$

Again, a dichotomy was shown in [BDW19] using the idea of Theorem 4.22. As fundamental codes, they used for C' a code for the arbitrarily varying cq-channel and for C'' a code for the arbitrarily varying wiretap cqq-channel, both reaching the capacity. If the transmission capacity for C'' is positive, one then gets as an identification capacity the transmission capacity of C''. The security follows by the strong security condition like in Theorem 4.22. Also the converse follows the same idea. As such, the following result is obtained.

Theorem 4.24 (Dichotomy) *Let $C(W)$ be the capacity of the arbitrarily varying cq-channel W and let $C_S(W, V)$ be the secure capacity of the arbitrarily varying wiretap cq-channel (W, V). Then,*

$$C_{SID}(W, V) = C_{SID}^{\text{sim}}(W, V) \tag{4.175}$$

$$= \begin{cases} C(W) & \text{if } C_S(W, V) > 0, \\ 0 & \text{if } C_S(W, V) = 0. \end{cases} \qquad \blacksquare \tag{4.176}$$

In this theorem the capacity is a single letter formula, but the condition if the capacity is positive is given by the multi-letter formula for the random coding secure capacity.

Remark 4.9 In the case of transmission it is possible to avoid the capacity being zero if the channel is *symmetrizable*, if we allow the sender and receiver to use common randomness. With this resource the capacity will not change if the channel is *non-symmetrizable*, but if the channel is *symmetrizable* then the capacity may go up from zero to the random coding capacity.

The situation appears different in the case of identification. We can, of course, use the same resource to get rid of the vanishing capacity in the *symmetrizable* case. However, note that a positive rate of common randomness, by the concatenated code construction of Ahlswede and Dueck [AD89a], increases the ID-capacity by the same amount. Fortunately, we are rescued by the fact that whenever common randomness is required to achieve the random coding capacity for transmission, then a rate of asymptotically zero is sufficient [Ahl78]. Thus, we could define random coding capacities with zero rate of common randomness without changing the notion for transmission, while obtaining a sound capacity concept for the identification problem.

Chapter 5
Quantum Error Correction

Active error correction is necessary to counter the effect of noise. Classically, it is necessary only for transmission and storage of data, because in these conditions noise has had a chance to accumulate over space or time. Otherwise, there is no active error correction in computation, because the signal levels are extremely high compared to the noise, thus giving extreme stability.[1]

Amplification, as we know it classically, cannot be used to achieve stable qubits because it cannot be done on quantum information without destroying it. This type of amplification hides an implicit measurement because for a classical signal to be amplified, its value must be known. $|\psi\rangle \rightarrow |\psi\rangle^{\otimes n}$ would give the desired properties of amplification for quantum states but is not a valid physical map because it is forbidden by the no-cloning theorem.[2] We will also see that using CNOTs to produce the map $|i\rangle \rightarrow |i\rangle^{\otimes n}$ only gives the quantum repetition code and does not amplify quantum information.

Additionally, in any quantum system, there is no distinction between the environmental noise in the quantum state and the one in the operations, because the latter just looks like noise in the Heisenberg picture rather than in the Schröedinger picture. Trapped ions are an example of qubits with long coherence time but with low interaction and slow operations, while superconducting qubits are an example

---

[1] Error correction is used in wireless and wired transmissions like WiFi, Bluetooth, HDMI, USB, and longer ranges and is used in storages like HDDs and SSDs, but also for shorter term storage like RAM memory in servers, and sometimes in cache. Otherwise for computation, like in processing units, error correction is not used, instead the data is decoded, computed, and re-encoded. If a hardware error happens, it is addressed at higher levels, and a personal computer can run years without a hardware error. In space applications, multiple systems are used to perform the same computation and provide redundancy against errors produced by cosmic rays, but the computation within these systems is not itself error corrected.

[2] $|\psi\rangle \rightarrow |\psi\rangle^{\otimes n}$ is clearly not linear because it contains an n-th power of the input.

© The Editor(s) (if applicable) and The Author(s), under exclusive license to
Springer Nature Switzerland AG 2021
R. Bassoli et al., *Quantum Communication Networks*, Foundations in Signal
Processing, Communications and Networking 23,
https://doi.org/10.1007/978-3-030-62938-0_5

of highly interacting qubits with fast gates, but also with short coherent times due to high noise. Therefore there is always a trade-off when designing a physical qubit because the qubits will interact with intended operations just as much as they interact with noise.

Therefore, in order to achieve the capabilities of quantum computation and quantum communication, quantum error correction is always needed, including during any processing. The landscape of quantum error correction is thus tasked with finding codes that allow for encoding and decoding not only of logical states, the analog of classical codewords, but also of logical gates, so that the operations are also error corrected. Such a property is known as fault-tolerance [Ren84, Sho96, Got98b, Pre98a].

In topological quantum computation, the physical qubits encode the information in physically delocalized degrees of freedom that require a global action to perform an operation [Kit03], with the goal of achieving more stable physical qubits than with other architectures. The reasonable assumption that noise is local provides higher stability and resilience to noise. However, the implementation of such qubit architecture is much more challenging than others, and the noise is not so low as to avoid the need for error correction [Pre98a]. Topological error correction codes are explicit code designs that exploit the same idea of requiring global operations on many physical qubits in order to perform logical operations. Both are an example of a topological order [Wen90].

5.1   Forward Error Correction Codes

In forward error correction, the decoder is not allowed to give feedback to the encoder. This avoids the latency of the feedback communication and is the only option for error correcting storage and processing, given the inability to communicate the feedback backward in time. The noise is modeled by a channel and can make distinct elements of the input set $\mathcal{A}$ indistinguishable at the output, generating errors at the decoder. Without considering randomization, a good encoder will thus map a set of messages into a subset of the inputs $\mathcal{C} \subseteq \mathcal{A}$, trying to minimize the probability of collisions at the decoder, which only depends on $\mathcal{C}$ and not on the actual mapping.

Any $\mathcal{C} \subseteq \mathcal{A}$ is generally called a (forward) error correction code, its size is the number of messages that can encode, and its elements are the codewords. For practical applications, both messages and codewords are often strings over the same alphabet $\mathcal{A}$. For encodings that map k-strings to n-strings, the code is known as a block code and is a subset $\mathcal{C} \subseteq \mathcal{A}^n$ of size $|\mathcal{C}| = |\mathcal{A}|^k$. If the alphabet is binary, then the codes are called binary codes.

In quantum communication, a good encoder will also allow a decoder to recover the sent state with high probability, but states that were only partially distinguishable at the sender will still be only partially distinguishable at the receiver. Thus the condition to satisfy is that all the recovered states $|\tilde{\psi}\rangle$ have good fidelity $|\langle \psi | \psi' \rangle| \sim$

1 with the original ones $|\psi\rangle$. Just like a classical encoder maps a classical system into a larger one, a quantum encoder will map a quantum system into a larger one. The equivalent of a deterministic encoder will be an isometry, and the encoded quantum states will still form a Hilbert space. In particular, the quantum error correction code will be a subspace $\mathcal{C}$, called the codespace or logical space of the Hilbert space $\mathcal{H}$ of the quantum system subject to the noise, often called the physical space. The dimension of the logical space $|\mathcal{C}|$ is the dimension of the encoded qudit. A quantum decoder, however, will map the larger physical Hilbert space into a smaller Hilbert space and thus *it will always be a noisy channel*.[3]

A point of departure from classical error correction is error detection. For classical codes, the information about the codewords is readily accessible, and therefore detecting whether a string contains an error can always be done by checking whether it is contained in the code. In quantum error correction, this information must be obtained with a measurement and constitutes the noisy part of decoding. Since $POVM$ measurements (see Chap. 2) are the result of partial information of a projective measurement outcome, we will consider only projective measurements without loss of generality.

To detect an error, we need to detect whether the quantum state is outside the logical space without affecting the valid states inside it. The measurement must leave the logical space intact, and thus it must contain the projector $P_{\mathcal{C}}$ onto $\mathcal{C}$. Furthermore, $P_{\mathcal{C}}$ must be one of the measurement projectors exactly, because otherwise, if the projector in the measurement is larger, there will be some error subspaces outside the logical space that will not be detected.

Any projective measurement corresponds to measuring an observable. In other words, there always exists an observable, such that measuring it distinguishes the logical space from its orthogonal subspace. The choice of this observable is not unique, but without loss of generality, we can assume that the eigenvalue of the logical space is 1. A possible example is the unitary $M = P_{\mathcal{C}} - (\mathbb{1} - P_{\mathcal{C}}) = 2P_{\mathcal{C}} - \mathbb{1}$. Thus we can always choose an observable for the code that is also a unitary operation on the physical space. The subspace with eigenvalue 1 of an operator M is called the stabilized subspace, or (improperly) the stabilizer of M, namely the subspace that is not changed by the action of the operator: $M |\psi\rangle = |\psi\rangle$.

In general, a quantum error correction code will define more than one observable to measure, and the collection of measurement outcomes is the error syndrome of the code; these observables are also called syndrome operators, often also referred to simply as syndromes, leaving the distinction clear from the context. Each syndrome will measure different subspaces containing $P_{\mathcal{C}}$ in order to identify the error in more detail and allow for correction. Still, each syndrome must leave the logical space intact, and therefore they must all commute with the logical space projector, which means that the logical space must itself be contained in the stabilized subspace of each syndrome operator. To detect all errors, the intersection of all stabilized

---

[3]However, there exists a version of each code with a noiseless decoder without mapping back to a smaller system [NC10, Box 10.1].

subspaces must exactly be the logical space, which brings us to a duality. *Any logical space can always be defined either explicitly or as the stabilized subspace of a set of operators.* Therefore the syndrome operators are also called stabilizer operators, or simply stabilizers, and the two terms can be used interchangeably, though in practice it is used only in the context of stabilizer codes introduced below. However, different syndromes can define the same logical space; thus, the correspondence between the two is not unique, as is already the case with the repetition code below. The definition as a stabilized subspace is generally more convenient for practical applications, while the error correction properties will only depend on the subspace itself.

If M is a syndrome operator, then M will commute with the logical space projector ($M P_{\mathcal{C}} = P_{\mathcal{C}} M$) and with any operator supported on it ($X = P_{\mathcal{C}} X P_{\mathcal{C}}$ implies $MX = XM$), in particular logical density matrices. A quantum error correction code correcting pure states will also correct mixed states.

Note 5.1 Because the Pauli matrices $\mathbb{1}$, X, Z, and iXZ are observables (Hermitian), unitaries, and an operator basis, they can play the role of syndrome measurements, gates, or errors. Which role they play should be clear from the context.

Additionally, since n-fold Pauli operators are the basis for the construction of relevant error correction codes, but at the same time usually appear as operators acting on few qubits, the quantum error correction literature uses a more compact notation than writing the n-fold tensor product. For any $i = 0, \ldots, n - 1$, each single-qubit Pauli operator is defined on the entire n qubits as

$$Z_i = \mathbb{1}^{\otimes i-1} \otimes Z \otimes \mathbb{1}^{\otimes n-i}, \qquad X_i = \mathbb{1}^{\otimes i-1} \otimes X \otimes \mathbb{1}^{\otimes n-i}. \tag{5.1}$$

Arbitrary Pauli operators are then obtained via simple multiplication, and Z_i and X_j obey the Pauli commutation relation if $i = j$, and commute otherwise. For example, in a space of four qubits, $X_1 Z_2 X_2 = \mathbb{1} \otimes X \otimes ZX \otimes \mathbb{1}$. The support of an n-fold Pauli operator is the set of qubits/indices where the operator does not act as the identity. In this example the support is $\{1, 2\}$.

As an example of a one-dimensional code (0 encoded qubits, the equivalent of a classical code with a single element), consider the subspace of the maximally entangled state $|\Phi\rangle = \frac{1}{\sqrt{2}}(|00\rangle + |11\rangle)$. This is the only state that satisfies $X^i Z^j \otimes X^i Z^j |\Phi\rangle$ for $i, j = 0, 1$. Thus the code can be specified either explicitly as this state or as the stabilized subspace of $X \otimes X$ and $Z \otimes Z$.

Like in classical applications, both the message and the encoded logical quantum states will be states over many identical qubits, or more generally over qudits with Hilbert space $\mathcal{H}$. A quantum block code will map k qudits to n qudits, the physical space will be $\mathcal{H}^{\otimes k}$, and the logical space will be a subspace $\mathcal{C} \subseteq (\mathcal{H})^{\otimes k}$ of dimension $|\mathcal{C}| = |\mathcal{H}|^k$. The previous example maps $k = 0$ qubits to $n = 2$ qubits.

5.2   Bit and Phase Errors: Quantum Repetition Code

The repetition code is the simplest imaginable code and consists simply of repeating n times the message $m \in \mathcal{A}$ into $\mathcal{A}^n$ ($k = 1$). For $n = 2$, the code can detect but not correct errors, and thus $n = 3$ is the simplest interesting example. For now we will focus on giving an example of a quantum code and will defer the discussion of *correctability* criteria to the next section. The repetition code is a linear code, meaning that the mapping from messages to codewords can be obtained as a linear map, and thus by matrix multiplication on the messages. Using the conventional left matrix multiplication, every codeword can be written as

$$c_m = m \cdot G \qquad\qquad G = \begin{pmatrix} 1 & 1 & 1 \end{pmatrix}, \qquad\qquad (5.2)$$

giving the codewords $c = 000, 111$ for $m = 0, 1$, and with the generalization to arbitrary n being simply a $1 \times n$ matrix G matrix filled with 1's.

Note 5.2 The linearity of this code must not be confused with the linearity of the quantum code. A non-linear classical code converted into a quantum code, as we will do below, will still create a quantum code that is a linear subspace. How the linearity of quantum codes translates to specific classes of quantum codes will be explained later.

The presence of an error can be detected computing the parities of two bits of the codeword, also called the syndromes for the classical code. This is again a linear process, and a minimum string of independent syndromes s can be computed as

$$s = c \cdot H \qquad\qquad H = \begin{pmatrix} 1 & 0 \\ 1 & 1 \\ 0 & 1 \end{pmatrix}. \qquad\qquad (5.3)$$

The parity checks are $c_1 \oplus c_2$ and $c_2 \oplus c_3$. If only one bitflip error occurs, the parity checks will detect it, and it can be corrected using a majority vote. Two bitflip errors will also be detected but will be indistinguishable from a logical NOT with a single bitflip, and thus a majority vote will produce a logical error.

A quantum repetition code is one of the simplest demonstrations that simply applying a classical error correction code will not give a fully error-correcting quantum code. We apply the repetition code map in superposition, which gives us the isometry:

$$|0'\rangle\langle 0| + |1'\rangle\langle 1| := |000\rangle\langle 0| + |111\rangle\langle 1|. \qquad\qquad (5.4)$$

The message standard basis $\{|0\rangle, |1\rangle\}$ is transformed into the logical standard basis $|0'\rangle = |000\rangle$ and $|1'\rangle = |111\rangle$. Every logical state in the logical space is written as $a|0'\rangle + b|1'\rangle = a|000\rangle + b|111\rangle$. The syndrome measurements are again parity measurements, this time defined by

$$Z \otimes Z \otimes \mathbb{1} \tag{5.5}$$

$$\mathbb{1} \otimes Z \otimes Z \tag{5.6}$$

$$Z \otimes \mathbb{1} \otimes Z, \tag{5.7}$$

which are all commuting and again redundant by one element, as for the classical case. Namely, the logical space is also (the intersection of) the subspace stabilized by any two of these parities. If a measurement outcome when $Z \otimes Z \otimes \mathbb{1}$ is 1, then the parity of the first two qubits is 0, otherwise, if the outcome is -1, then the parity is 1. Alternatively, they can all be transformed into an observable with the correct outcome via, e.g., $\frac{1}{2}(\mathbb{1} - Z \otimes Z \otimes \mathbb{1})$.

Note 5.3 Measuring the parities is not the same as measuring each Z separately and then computing the parity because the projectors are different. Namely, measuring $Z \otimes \mathbb{1}$ and $\mathbb{1} \otimes Z$ in sequence is equivalent to measuring with projectors $|00\rangle\langle00|$, $|11\rangle\langle11|$, $|01\rangle\langle01|$, and $|10\rangle\langle10|$. However, the measurement of $Z \otimes Z$ has projectors $|00\rangle\langle00| + |11\rangle\langle11|$ and $|01\rangle\langle01| + |10\rangle\langle10|$. It is a straightforward check to verify that the first measurement disturbs the codespace, while the latter does not.

In general, any projective measurement with d projectors must be equivalent to a measurement on a single d-dimensional qudit. For parity checks, which are the tensor products of Pauli Z on the interested qubits, this is done by computing the parity coherently with CNOTs on an ancilla qubit and measuring the ancilla. For $Z \otimes Z$, this is

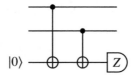

and can be generalized to an arbitrary tensor product of Pauli Z. Similarly, if we need to measure conjugate parities like $X \otimes X$, simply perform the same circuit in the conjugate basis, namely conjugate the circuit with Hadamard rotations.

A bit of squinting should convince the reader that in order to correct a correctable error, the original information must still be contained in the collapsed state with an error. Namely, if the logical space has dimension d (in our case two), then the subspace of each error syndrome (the collection of measurement outcomes) must be at least of dimension d. Furthermore, if the correction operation is noiseless, then the error syndrome subspaces must exactly be of dimension d. This is indeed what also happens in this repetition code. Any input state is collapsed into any of the four two-dimensional subspaces as shown in Table 5.1. The parities allow us to correct one bitflip error using a majority vote, as is also shown in the table and done in the classical case.

As previously mentioned, we need to be able to perform an operation on the logical qubit without exiting the encoding. Such operations must commute with

Table 5.1 Three qubit error correction for bitflip errors

Subspace	Syndrome	Correction
$\{\lvert000\rangle,\lvert111\rangle\}$	$+1,+1,+1$	$\mathbb{1}\otimes\mathbb{1}\otimes\mathbb{1}$
$\{\lvert100\rangle,\lvert011\rangle\}$	$-1,+1,-1$	$X\otimes\mathbb{1}\otimes\mathbb{1}$
$\{\lvert010\rangle,\lvert101\rangle\}$	$-1,-1,+1$	$\mathbb{1}\otimes X\otimes\mathbb{1}$
$\{\lvert001\rangle,\lvert110\rangle\}$	$+1,-1,-1$	$\mathbb{1}\otimes\mathbb{1}\otimes X$

the syndrome operators in order to allow the recovery from errors. We make the example of the logical Pauli gates. A logical phaseflip is any operator that acts as the Pauli Z on the logical space, $P_{\mathcal{C}}Z'P_{\mathcal{C}} = \lvert0'\rangle\langle0'\rvert - \lvert1'\rangle\langle1'\rvert$, *and* commutes with the syndromes. Any of $Z\otimes\mathbb{1}\otimes\mathbb{1}$, $\mathbb{1}\otimes Z\otimes\mathbb{1}$, and $\mathbb{1}\otimes\mathbb{1}\otimes Z$ is thus a logical phaseflip.

Similarly, a logical bitflip is any operator that acts as the Pauli X on the logical space, $P_{\mathcal{C}}X'P_{\mathcal{C}} = \lvert0'\rangle\langle1'\rvert + \lvert1'\rangle\langle0'\rvert$, and commutes with the syndromes. As for the classical case, a valid choice is the bitflip on each qubit $X' = X\otimes X\otimes X$, which commutes with the syndromes, because with each of them, it overlaps with their Z operators in exactly two qubits (an even number of commutations of X and Z produces an even number of -1's, and thus bitflips and phaseflips that overlap on an even number of qubits always commute).

The commutativity condition with the syndromes further means that the logical gates are also logical gates in the syndrome subspaces. Each of the subspaces in Table 5.1 could be chosen as the logical subspace, and the logical operators would not change, only the role of the syndrome of the correction operators would change.

The quantum repetition code can correct one bitflip error but is not considered a code that can correct one quantum error. In particular, a single phaseflip error produces a logical error and thus cannot be corrected. As they commute with the syndromes, they cannot be detected. In other words, since the logical phaseflip operators are the same as single phaseflip errors, they are indistinguishable and thus uncorrectable.

The code can be changed into correcting phaseflips by exchanging the roles of X and Z. The encoding simply requires an additional Hadamard gate on each qubit, and the logical subspace becomes $\{\lvert+++\rangle,\lvert---\rangle\}$. Three phaseflips will act as a logical bitflip, while any bitflip will act as a logical phaseflip; thus in this case, bitflips are the ones that cannot be corrected. In short, a repetition code can be made to correct either bitflips or phaseflips but not both at the same time, and thus they cannot correct an arbitrary Pauli error.

5.3 Single Pauli Error: Shor's Error Correction Code

We can correct a single arbitrary Pauli error by concatenating a repetition code for phaseflips and a repetition code for bitflips; the result is Shor's error correction code displayed in Fig. 5.1. The order of the two codes is arbitrary, but by correcting the phaseflips first, we can save six Hadamard gates. Thus this is the usual choice.

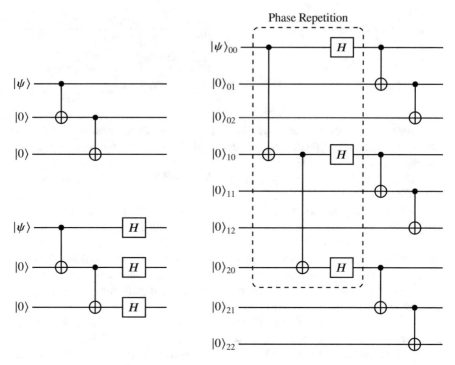

Fig. 5.1 Encoding circuits for the bitflip and phaseflip repetition codes (left) and for Shor's error correction code (right). For the latter, first a quantum repetition code against the phaseflip is encoded (dotted box). Then each qubit is encoded against the bitflips, each with another repetition code

First we fix the notation for the encoded qubits. Given a state $|\psi\rangle$, we denote by $|\psi'\rangle$ the logical state encoded by a bitflip-correcting repetition code. In particular we have

$$|i'\rangle = |iii\rangle \qquad i = 0, 1 \qquad |\pm'\rangle = \frac{1}{\sqrt{2}}(|0'\rangle \pm |1'\rangle) = \frac{1}{\sqrt{2}}(|000\rangle \pm |111\rangle).$$

$$(5.8)$$

Each code maps into three qubits, and thus we need nine qubits that we will index with two indices $i, j = 0, 1, 2$ in order to use the compact notation for the Pauli operators. Let $|\psi\rangle = c_0 |0\rangle + c_1 |1\rangle$ be the state to encode in qubit 00. We will use qubits 10 and 20 to encode the first phaseflip-correcting repetition code. The encoded state becomes

$$c_0 |+++\rangle + c_1 |---\rangle \qquad\qquad (5.9)$$

on qubits 00, 10, and 20. The phase syndromes are $X_{00}X_{10}$ and $X_{10}X_{20}$, and the logical phaseflips are $Z'_0 := Z_{00}Z_{10}Z_{20}$, X_{00}, X_{10}, and X_{20}.

At this point, we encode each qubit above with the bitflip-correcting repetition, which maps $|\pm\rangle \rightarrow |\pm'\rangle$. For each qubit $i0$, we use qubits $i0$, $i1$, and $i2$, and we denote with $|\pm'_i\rangle$ the encoded state. The final encoded state is

$$c_0 |+'_0\rangle |+'_1\rangle |+'_2\rangle + c_1 |-'_0\rangle |-'_1\rangle |-'_2\rangle . \tag{5.10}$$

As possible logical phaseflips, we have, for example, $X'_0 = X_{00}X_{01}X_{02}$, $X'_1 = X_{10}X_{11}X_{12}$ or $X'_2 = X_{20}X_{21}X_{22}$. Each will map $|\pm'_i\rangle \rightarrow \pm|\pm'_i\rangle$, thus performing the encoded phaseflip $|\pm' \pm' \pm'\rangle \rightarrow \pm|\pm' \pm' \pm'\rangle$.

As possible logical bitflips, we have, for example, $Z'_0 = Z_{00}Z_{10}Z_{20}$, $Z'_1 = Z_{01}Z_{11}Z_{21}$ or $Z'_2 = Z_{02}Z_{12}Z_{22}$. Each Z_{ij} will map $|\pm'_i\rangle \rightarrow |\mp'_i\rangle$, thus acting here as a bitflip, and we need one on each i to perform the encoded bitflip $|\pm' \pm' \pm'\rangle \rightarrow |\mp' \mp' \mp'\rangle$.

The set of measurements $\{Z_{00}Z_{01}, Z_{01}Z_{02}\}$ is used for the error syndromes to detect a bitflip of the first, second, and third qubits. Likewise, $\{Z_{10}Z_{11}, Z_{11}Z_{12}\}$ give the error syndromes to detect a bitflip of the fourth, fifth, and sixth qubits, and $\{Z_{20}Z_{21}, Z_{21}Z_{22}\}$ to detect a bitflip of the seventh, eighth, and ninth qubits. A phaseflip is detected by the set of measurements $\{X'_0X'_1, X'_1X'_2\}$.

The set of bitflips $\{X_{ij} : j = 0, 1, 2\}$ is corrected independently for different i; thus, up to three bitflip errors can be corrected simultaneously as long as they belong to *different* sets, namely if they act on different i's. The set of phaseflips $\{Z_{ij} : j = 0, 1, 2\}$ leads to the same error; therefore, up to three phaseflip errors can be corrected simultaneously as long as they belong to the *same* set, namely if they act on the same i.

The syndrome measurements will collapse all the single-qubit Kraus operators into a mixture of Pauli Kraus operators and thus of phaseflips and bitflips. The intersection of the errors that are correctable simultaneously only allows one qubit to be corrected from both a phaseflip and a bitflip. Therefore only one general error can be corrected with this code. More precisely, for any single-qubit quantum channel Λ, this code will correct the output $\Lambda(\rho)$ back to ρ. However, it will fail to correct the quantum channel noise in general if it is a channel of more than one qubit.

5.4   Error Correction Condition and Code Distance

So far we have focused only on the correction of the discrete set of Pauli errors. However, quantum errors are continuous and most generally defined as the action of a quantum channel. Correcting a quantum channel is equivalent to correcting its Kraus operators, and thus we can restrict our focus to correcting a single generic Kraus *error* E (satisfying $0 \leq E^\dagger E \leq \mathbb{1}$). We thus need error correction codes that can correct against all these errors, and more generally, we need a condition to verify this requirement.

It turns out that correcting E is not more difficult than correcting Pauli errors, and that Shor's error correction code indeed corrects any error map on a single qubit. Heuristically, the reason is that even though the actual error might be E, the syndrome measurements collapse any Kraus error into a mixture of Pauli errors

$$E\rho E \rightarrow \sum_{i,j=0,1} k_{ij}(X^i Z^j)\rho(X^i Z^j)^\dagger, \qquad (5.11)$$

for which we omit the derivation. This is common in practical error correction codes: arbitrary single-qubit errors can be reduced to correcting a discrete set of errors because the measurement collapses the continuous set into the discrete one and it has a formal explanation in the quantum error correction condition. In general though, as we mentioned before, the choice of syndrome operators is not unique, so while the logical space might be defined by the syndromes, the set of errors that can be corrected is a property of the logical space itself. Therefore the intuition for the quantum error correction condition cannot depend on the syndromes like the one we have just given.

Classically, the error correction condition is simple, a set of errors is correctable if it maps different codewords into disjoint subsets. Quantum-mechanically, different logical states that are not orthogonal cannot be mapped to orthogonal states, but we can impose this condition for a basis of logical states.

Let $\{ E_i \}$ be a set of Kraus operators. To be able to correct this set, we at least need to require that for the different logical basis states $|i'\rangle$ and $|j'\rangle$, the produced error states are all orthogonal. More precisely, for any $i \neq j$ and any E_k and E_l we need to require $\langle i'| E_k^\dagger E_l |j'\rangle = 0$, leaving us to determine $\langle i'| E_k^\dagger E_l |i'\rangle = C_{kli}$. Similarly, we should require that any two orthogonal logical states, and not just the logical basis states, remain orthogonal. This simplifies the condition further, imposing that C_{kli} should not depend on i. This is because otherwise the Kraus error acting on a superposition of logical state would change the coefficient in the superposition, in which case we can construct orthogonal states that become non-orthogonal after the action of the Kraus error. Therefore C_{kli} cannot depend on i, and we can write the quantum error correction condition as

$$\langle i'| E_k^\dagger E_l |j'\rangle = C_{kl}\delta_{ij}, \qquad (5.12)$$

for some Hermitian matrix C. It turns out that this is the necessary and sufficient condition [BDSW96, KLV00]. We can tensor and sum the logical matrix basis $(|i'\rangle\langle j'|)$ and rewrite the condition directly in terms of the logical space projector:

$$P_{\mathcal{C}} E_k^\dagger E_l P_{\mathcal{C}} = C_{kl} P_{\mathcal{C}}. \qquad (5.13)$$

With $V = \sum_k \langle k| \otimes E_k,$[4] we can further rewrite it as

$$(\mathbb{1} \otimes P_{\mathcal{C}}) V^\dagger V (\mathbb{1} \otimes P_{\mathcal{C}}) = C \otimes P_{\mathcal{C}}. \qquad (5.14)$$

Equations (5.12) to (5.14) are all equivalent versions of the quantum error correction condition.

Recall that the choice of Kraus operators of a quantum channel is not unique. Changing the choice of Kraus errors equals a unitary conjugation on C. Since C is Hermitian, if Eq. (5.14) is satisfied, there always exists an equivalent choice of Kraus errors such that C is diagonal. A diagonal C corresponds to a set of errors that is not degenerate for the code $\mathcal{C}$, namely, each different Kraus error maps to a different syndrome. However, this cannot be the general case as shown by Shor's error correction code; there some sets of phaseflips lead to the same error, and thus C is not diagonal for single-qubit Pauli errors.

The reduction to Pauli errors finally emerges by noting that if a set of errors satisfies Eq. (5.13), then the whole linear space spanned by these errors also satisfies it. Since k-fold Pauli operators are a basis for the linear space of operators on k qudits, being able to correct them implies being able to correct any k-qudits Kraus operators, and thus any error quantum channel.

Code Distance

In practical applications it is often reasonable to assume that the error is identical and memoryless over a certain alphabet. One of the achievements of information theory is the realization that in the limit of many identical uses of a channel, the action becomes *typical*, meaning that the number of appearances of each error in n uses of the channel will converge to a certain ratio, but otherwise will be uniformly distributed over the n uses. When using an error correction code in these scenarios, it thus becomes irrelevant where the error can be corrected, and it is only important how many errors can be corrected in arbitrary positions. Let n_e be the number of correctable errors. Namely, let us consider the codes such that if any number $n = 0, \ldots, n_e$ of bits/qubits suffer an error, then there is a decoding that corrects any state in the codespace.

Classically, the set of n_e errors takes each codeword and maps it into a subset of possible strings. This is the error ball of distance n_e under the Hamming distance defined as the map $d^H : \mathcal{A}^n \times \mathcal{A}^n \to \{0, \cdots, n\}$ that counts the different symbols: $d^H(x, y) = |\{i \in \{1, \cdots, n\} : x_i \neq y_i\}|$. For binary codes ($\mathcal{A} = \{0, 1\}$), the Hamming distance equals $d^H(x, y) = w(x \oplus y)$, where w is the Hamming weight $w(x) = |\{i : x_i \neq 0\}|$.

A block code is said to have distance d if $d^H(x, y) \geq d$ for any $x, y \in \mathcal{C}$. At the same time, a code can correct n_e errors if the balls of any two codewords are disjoint. If all the codewords have Hamming distances above $2n_e$, then all the error

---

[4]This is not exactly the isometry purifying a channel because it differs by a partial transpose on the environment.

terms with n_e errors or fewer cannot generate a collision and those error terms can be corrected with certainty.

Quantum-mechanically, we just need to appropriately replace the condition and consider only the Pauli errors. The weight of an n-fold Pauli operator is defined as the size of its support. The *distance* of a quantum error correction code in n physical qubits/qudits is defined as the maximum weight d, such that for all Pauli errors $E \neq \mathbb{1}$ of weight d or less, it holds

$$P_{\mathcal{C}} E P_{\mathcal{C}} = C_E P_{\mathcal{C}}, \tag{5.15}$$

where C_E is a complex number depending on the error which will be zero for non-degenerate codes. Since $E^\dagger$ is also a Pauli error for all Pauli errors E, and the product of two Pauli errors of weight n_e has weight at most $2n_e$, just like in the classical case, Eqs. (5.13) and (5.15) together give that a quantum code with distance above $2n_e$ can correct n_e errors.

Sometimes the notation $[\![n, k, d]\!]_{\mathcal{H}}$ is used to denote quantum block codes from k to n qudits in $\mathcal{H}$ with distance d[Pre98b], even though the linearity of classical linear block codes does not play the same role as the linearity of quantum codes. Here we use the convention [NC10, Got97] to reserve the notation for stabilizer codes. However, the need for fault-tolerance has rendered the code distance, an inappropriate measure for quantum error correction. For example, in order to achieve other desirable properties, families of topological codes of increasing number of physical qubits n are able to correct most errors with weight of order n, thus displaying an "effective" distance also growing with n, but there will be few logical errors with weight sublinear in n that will make the code distance grow much slower.

5.5 Linear Codes and Stabilizer Codes

The example of the repetition and Shor's error correction codes display a general way of constructing quantum error correction codes. Any classical error correction code $\mathcal{C}$ can define a quantum error correction code with the logical basis $\{|c\rangle : c \in \mathcal{C}\}$. While these type of codes will only be able to correct bitflip errors, Hadamard gates can be used to convert them into only phaseflip-correcting codes, and the two types can be combined into full quantum error-correcting codes.

In this section we focus on another aspect displayed by these codes, namely that classical linear codes define quantum error-correcting codes with syndromes that are all Pauli operators.

5.5.1 Linear Block Codes

A linear code from k-strings to n-strings is a classical block code $\mathcal{C} \subseteq \mathcal{A}^n$ such that $\mathcal{C}$ is also a k-dimensional linear subspace of $\mathcal{A}^n$, where $\mathcal{A}$ must now be a finite number field. $[n, k]_{\mathcal{A}}$ denotes the set of such codes and $[n, k, d]_{\mathcal{A}}$ the subset with code distance d. If $\mathcal{A}$ is omitted, the alphabet is assumed to be binary. We remark again that the linearity of such codes is different than the linearity of quantum error correction codes: if we form a quantum code from any classical code (as done for the bit-repetition code), the result will always be a subspace of the physical Hilbert space.

That said, as for the repetition code, by linearity, the encoding of a linear code $\mathcal{C}$ can always be written as the left matrix multiplication $m \cdot G$ of the message $m \in \mathcal{A}^k$ with a non-unique $k \times n$ matrix G, referred to as a generator matrix.

Every subspace $\mathcal{C}$ has an orthogonal subspace $\mathcal{C}^{\perp}$ that equivalently identifies the codewords $c \in \mathcal{C}$ as the orthogonal vectors to $\mathcal{C}^{\perp}$. This is done verifying the system of equations $c \cdot H^{\mathsf{T}} = 0$, where H is a non-unique $(n - k) \times n$ matrix called the *check matrix* of $\mathcal{C}$ (parity check matrix if the alphabet is binary). The two matrices G and H are dual to each other, meaning that $\mathcal{C}^{\perp}$ also constitutes a valid $[n, n - k]$ linear code known as the dual code of $\mathcal{C}$, with H being a valid generator matrix and G a valid parity check matrix of $\mathcal{C}^{\perp}$. The checks, or syndromes, are the symbols of $c \cdot H^{\mathsf{T}}$, which detect an error whenever they are non-zero.

Linear codes allow for much easier implementation of error correction than arbitrary codes do, with the generator matrix suitable for efficient encoding and the check matrix suitable for error detection.

5.5.2 Stabilizer Codes

For simplicity, from now on we will restrict the classical and quantum alphabets to bits and qubits. Consider the quantum error correction code $\{|mG\rangle\}_{m \in \mathbb{Z}_2^k}$ spanned by the codeword of a classical binary linear code with generator matrix G, as is done for the repetition code. We can now see how the linearity of G is different from the linearity of the logical subspace. Indeed, the generator matrix in the encoder $|m\rangle \rightarrow |mG\rangle$ acts inside the ket, thus only as a function on the indices and not as a linear operator G on the logical space, which would act as $|m\rangle \rightarrow G|m\rangle$.

We can recover the use of the classical linearity with the syndrome operators defined by the parity check matrix. If we denote with $h \in H$ the n-strings that are rows of H, then we can define the syndrome operators

$$Z^h = Z^{h_1} \otimes \cdots \otimes Z^{h_n}. \tag{5.16}$$

The parity check condition $c \cdot h = 0$ implies that if c is a Codeword, then $|c\rangle$ is stabilized by all Z^h, and thus $|c\rangle$ is a logical basis state, and otherwise $|c\rangle$ will have the eigenvalue -1 for some syndrome Z^h.

The set of syndromes $\{Z^h\}_{h \in H}$ exactly identifies the subspace of $\{|c\rangle\}_{c \in \mathcal{C}}$, the logical space, as the stabilized subspace. Furthermore, $\{Z^h\}_{h \in H}$ generates an Abelian representation of the dual subspace $\mathcal{C}^{\perp}$, namely the group generated by $\{Z^h\}_{h \in H}$ via multiplications is commutative and in a one-to-one correspondence with the subspace spanned by H, as is shown by $(Z^{h_1})^{\alpha_1}(Z^{h_2})^{\alpha_1} = Z^{\alpha_1 h_1 + \alpha_2 h_2}$. The linearity of the classical codes translates into the syndromes forming a subgroup of the Pauli group, known as the stabilizer subgroup of the code. While the parity check matrix can vary, and thus also the set of stabilizer generators, the dual subspace and stabilizer subgroup do not.

If instead we want to correct phaseflip errors, we will simply rotate the code with $H^{\otimes n}$. The syndromes then become $\{X^h\}_{h \in H}$ with

$$X^h = X^{h_1} \otimes \cdots \otimes X^{h_n}. \tag{5.17}$$

However, we have seen that a code correcting at least a full set of single-qubit Pauli errors will not have only phaseflips or only bitflips in the stabilizer group, thus justifying the following definition.

A *stabilizer code* over n is a quantum error correction code defined as the stabilized subspace of an Abelian (so that the syndromes commute) subgroup S of the n-qubit Pauli group $\mathcal{P}_n$ known as the stabilizer. The code encodes k qubits if S has $n - k$ independent generators (each independent generator divides the space in half). $[\![n, k]\!]$ denotes the set of stabilizer codes from k to n qubits and $[\![n, k, d]\!]$ the ones with distance d.

The Pauli operators in S have no effect on the code as gates, the ones outside are errors, and the errors that do not commute with S are detectable errors, while the ones that do are undetectable ones.

In the example of the repetition code, the stabilizer is $\{\mathbb{1}, Z_1 Z_2, Z_2 Z_3, Z_1 Z_3\}$, which is generated by either $\{Z_1 Z_2, Z_2 Z_3\}$, $\{Z_1 Z_2, Z_1 Z_3\}$, or $\{Z_2 Z_3, Z_1 Z_3\}$. The single-qubit phaseflips together with $X_1 X_2 X_3$ generate the undetectable errors, the rest are detectable errors.

A reason why stabilizer codes are desirable in practice is their interplay with universal gate sets. We have mentioned in Sect. 3 that the Clifford group is a set of quantum gates containing the Pauli group that can be efficiently simulated. With stabilizer codes, the evolution of errors across Clifford circuits can be efficiently tracked and the syndromes accumulated, delaying the correction to a single operation at the end, thus reducing the number of executed gates and induced noise errors.

5.5.3 Calderbank–Shor–Steane (CSS) Codes

In practice, it is also convenient to construct codes with generators that can be written as either only phaseflip or only bitflip operators. This corresponds to having two parity check matrices H_Z and H_X generating the syndrome operators and correcting, respectively, bitflips and phaseflips independently. The commutation condition then corresponds to H_Z and H_X being orthogonal, namely all the phaseflip syndromes must always overlap on an even number of qubits with the bitflip syndromes. The first examples of such codes were CSS codes.

Calderbank, Shor, and Steane (CSS) showed how to choose two classical linear codes appropriately in order to define a stabilizer code using co-set states. Historically, CSS codes have been the precursors of the stabilizer formalism and stabilizer codes. They provide a standard way to transfer the intuition and knowledge of classical linear codes into quantum error correction and are thus an essential tool to cover. However, there are advantages and insights to gain in constructing native stabilizer codes that are not based on classical intuition, as we will see with topological codes.

Let $\mathcal{C}_1$ be an $[n, k_1, d]$ (binary) code. Let $\mathcal{C}_2$ be an $[n, k_2 < k_1]$ subcode (thus a subspace) of $\mathcal{C}_1$, such that the dual code $\mathcal{C}_2^\perp$ is an $[n, n - k_2, d]$ code. Let H be a parity check matrix for $\mathcal{C}_1$ and G a generator matrix for $\mathcal{C}_2$. In particular these matrices will be orthogonal, $G \cdot H^\mathsf{T} = 0$, because G generates a subspace of $\mathcal{C}_1$. The co-set space $\mathcal{C}_1/\mathcal{C}_2$ itself forms an $[n, k_1 - k_2]$ linear block code.

The $CSS(\mathcal{C}_1, \mathcal{C}_2)$ code is defined as the code with stabilizer generators

$$\{Z^h\}_{h \in H} \cup \{X^g\}_{g \in G}, \tag{5.18}$$

which are indeed commuting, because $G \cdot H^\mathsf{T} = 0$ implies $h \cdot g = 0$, which makes phaseflip and bitflip operators always overlap on an even number of qubits and makes them commute. The number of stabilizers is $n - k_1$ from $\mathcal{C}_1$ and k_2 from $\mathcal{C}_2$, and thus the number of encoded qubits is $k_1 - k_2$, the number of bits encoded by $\mathcal{C}_1/\mathcal{C}_2$.

Traditionally, $CSS(\mathcal{C}_1, \mathcal{C}_2)$ explicitly defines the logical basis as the co-set states of $\mathcal{C}_1/\mathcal{C}_2$:

$$|c + \mathcal{C}_2\rangle := \frac{1}{\sqrt{|\mathcal{C}_2|}} \sum_{a \in \mathcal{C}_2} |c + a\rangle \qquad \forall c + \mathcal{C}_2 \in \mathcal{C}_1/\mathcal{C}_2. \tag{5.19}$$

We can check that these are indeed the logical states of $\{Z^h\}_{h \in H} \cup \{X^g\}_{g \in G}$ by checking that for any error bitstrings x and z, the syndromes of the affected codestate $X^x Z^z |c + \mathcal{C}_2\rangle$ are given by

$$Z^h X^x Z^z |c + \mathcal{C}_2\rangle = (-1)^{h \cdot x} X^x Z^z |c + \mathcal{C}_2\rangle, \tag{5.20}$$

$$X^g X^x Z^z |c + \mathcal{C}_2\rangle = (-1)^{z \cdot g} X^x Z^z |c + \mathcal{C}_2\rangle. \tag{5.21}$$

Thus, measuring all the phaseflip syndromes reveals the error syndrome $x \cdot H$ of $\mathcal{C}_1$ and measuring all the bitflip syndromes reveals the error syndromes $z \cdot G$. More explicitly, any error detectable by $\mathcal{C}_1$ will also be detectable as a bitflip error by the phaseflip stabilizers, and any error detectable by $\mathcal{C}_2^\perp$ will be detectable as a phaseflip error by the bitflip stabilizers.

Since $\mathcal{C}_1$ and $\mathcal{C}_2$ can detect arbitrary errors of weight d, the resulting $CSS(\mathcal{C}_1, \mathcal{C}_2)$ will detect arbitrary Pauli errors of weight d. More generally, if $\mathcal{C}_1$ has distance d_1 and $\mathcal{C}_2$ has distance d_2, then the distance of $CSS(\mathcal{C}_1, \mathcal{C}_2)$ is $\min(d_1, d_2)$.

5.6   Universal Logical-Gate Sets

One of the reasons for not using CSS codes and moving toward native stabilizer code constructions is the need to find a universal set of logical gates. As mentioned before, once quantum information is encoded, it will often not be possible to decode and make the quantum computation non-error corrected, because even during computation time the noise levels will be too high. It is thus necessary to find out how to implement the logical Pauli gates, logical Hadamards, logical CNOTs, etc. on the encoded states and to perform arbitrary error corrected computation. It will also be necessary for these logical gates to form a universal gate set.

Consider for a moment the quantum repetition code. While finding logical Pauli matrices is relatively simple ($X_1 X_2 X_3$ and Z_1), implementing a logical Hadamard requires implementing $|iii\rangle \rightarrow |000\rangle + (-1)^i |000\rangle$, which as an example is not implemented by $H^{\otimes 3}$ and might require complex compilation. Classical linear codes are optimized for efficient encoding and decoding. However, quantum stabilizer codes also need to be optimized for efficient logical computation.

Another problem created by complicated compilation of logical gates is error propagation. Suppose that CNOT gates are needed to implement a logical Hadamard on an $[[n, 1]]$ code. Because $\text{CNOT} X \otimes \mathbb{1} = X \otimes X \text{CNOT}$ and $\text{CNOT} \mathbb{1} \otimes Z = Z \otimes Z \text{CNOT}$, a CNOT can propagate and increase the weight of an error making it uncorrectable or triggering a logical error. At the very least, the errors would need to be backtracked to find where they actually came from.

Transversality
Transversality is a desirable property of quantum block codes aimed at reducing the complexity of a gate (in particular if locality is required in the codes) and error propagation. A logical gate U' is *transversal* if it can be implemented as $V^{\otimes n}$ for some physical gate V on the physical qubits. In the case of a logical CNOT that needs to act on two logical qubits, transversality would be achieved if it could be implemented with physical CNOT gates only between the qubits of the two codes and not among the qubits of the same code. If a logical gate is transversal, then the errors can propagate between codes where they can be corrected, but they will not

Fig. 5.2 The transversality of the logical CNOT (boxed) in the repetition code. Note that there are two instances of the repetition code in order to have a logical CNOT

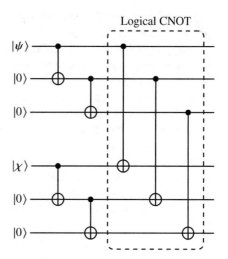

propagate and multiply within a single code where they can do damage. The logical CNOT in the repetition code is transversal as illustrated in Fig. 5.2.

However, there are trade-offs in the search for a universal set of transversal gates. The Eastin–Knill theorem [EK09] shows that there exists no single code with a universal set of transversal gates. The work-around for this no-go theorem is alternating the use of a stabilizer code and a stabilizer subcode with complementary transversal gate sets [Bom15]. The stabilizer of the code will be a subgroup of the stabilizer of the subcode. Alternating the use of the two stabilizer groups enables the two codes to alternate with each other and use their set of transversal gates on the same quantum information, thus allowing us to achieve a universal logical-gate set made of transversal gates.

5.7 Topological Stabilizer Codes

Topological stabilizer codes use the topological properties of manifolds by arranging the qubits in graphs embedded on the manifolds. Of particular interest are surface manifolds, like the torus, which we will take as an exemplary case, because physical qubits on a chip will be arranged on surfaces. The goal is to protect against unstructured noise by encoding the qubits in structured degrees of freedom that require coordinated global action in order to be changed. In order to achieve these properties, topological error correction codes combine the concepts of qubit representations, code locality, and homology, for which we give a quick overview.

Qubit Representations
We have defined a qubit as any two-dimensional Hilbert space with arbitrary unitary operations and arbitrary Hermitian observables. The Pauli matrices are particular operations and observables, but most importantly, they are a basis for

both of them in two-dimensional complex space. They are a linear basis for the Hermitian observables, since any Hermitian matrix M can be written as $M = \sum_{k,l=0,1} m_{kl}(i^{kl}X^kZ^l)$, and thus knowing their expectation values implies knowing the expectation value of any observable. They are also a basis for the unitary matrices via exponentiation, since any unitary matrix U can be written as the exponential of a Hermitian matrix H as e^{iM}. The abstract algebra of the 2×2 Hermitian matrices is called $u(2)$, while the abstract group of 2×2 unitary matrices is called $U(2)$, which is generated by $u(2)$ via exponentiation. In general, the state space of a qubit does not have to be $\mathbb{C}^2$, and the operator/observable basis does not have to be the 2×2 Pauli matrices. Indeed, in quantum error correction, the goal is to find a qubit embedded in larger Hilbert spaces, with qubit operations now being a subgroup of the larger unitary group (a group representation of $U(2)$), and the qubit observables being a subalgebra of the Hermitian matrices (an algebra representation of $u(2)$).

A qubit is most generally defined being as such a representation of $U(2)$, with the quantum error correction codes seen so far being an example. The result to be exploited from representation theory is that we do not need to know the underlying vector space in order to prove that we have a qubit, and we only need to prove that we have found an algebra of operators that have the same structure as the Pauli matrices. More precisely, we have a qubit any time we can find two operators $\bar{X}$ and $\bar{Z}$ such that they square to the identity $\bar{X}^2 = \bar{Z}^2 = \mathbb{1}$, and anti-commute $\bar{X}\bar{Z} = -\bar{Z}\bar{X}$, which we can write concisely as $\{\bar{X}, \bar{Z}\} = 0$ with $\{A, B\} := AB + BA$ known as the commutator or Poisson bracket of A and B.[5] The generalization to k qubits $\bar{X}_1, \bar{Z}_1, \ldots \bar{X}_k, \bar{Z}_k$ is also straightforward: $\bar{X}_i$ and $\bar{Z}_j$ must anti-commute for the same qubit $i = j$ and commute among different qubits.

In the example of the repetition code, these conditions are satisfied by, for example, $\bar{X} = X_1X_2X_3$ and $\bar{Z} = Z_1$, and in the case of Shor's code by, for example, $\bar{X} = X_{00}X_{01}X_{02}$ and $\bar{Z} = Z_{00}Z_{10}Z_{20}$.

Local Codes

A stabilizer code is local if all the qubits are organized in a graph and all the stabilizer generators have bounded weight and the support is bounded in the graph (contained in a ball of fixed radius). In particular, the graph will often represent a real-world configuration and as such, it will be an embedded graph on a D-dimensional surface such that no two edges intersect each other. The edges represent the available native two qubit gates, and the bounded support represents ancilla qubits used to compute the parities, which are only able to interact with physical qubits close to it.

Local stabilizer codes are particularly appealing for solid state architectures (thus memories), and more generally, when full connectivity between the qubits is not

---

[5] $\{\bar{X}, \bar{Z}\} = 0$ and $\bar{X}^2 = \bar{Z}^2 = \mathbb{1}$ imply $[\bar{X}, \bar{Z}] = -2i(i\bar{X}\bar{Z})$, where $[A, B] := AB - BA$ is the commutator or Lie bracket of A and B, which is the standard way of writing the structure constants for $su(2)$.

available. As we have seen, CNOT gates are needed to implement parity checks, and it may become costly if the involved qubits are fixed far from each other or need to be moved around. Compiling long-distance gates for error correction increases the measuring/decoding time and accumulates errors.

For example, in trapped-ions architectures, qubits (the ions) will be organized in various small chains, since large chains become unstable. Inside the chains, vibrational modes can be used to make the qubits interact, but to interact among chains, qubits must be shuttled around. In solid state chips, qubits cannot move and are fixed on a surface together with their connectivity. Some architectures may have a few interaction paths that intersect on the surface, but mostly the interactions are reduced to nearest neighbors and full connectivity will be impossible

In the repetition code, if we assume that the qubits are fixed in a line like in the circuit diagram (Fig. 5.1), then the local set of generators with only nearest-neighbor interactions is $\{Z_1 Z_2, Z_2 Z_3\}$ because qubits one and three are not nearest neighbors.

Homology Classes

An error on a physical qubit in a local code is a local transformation. Since we want to protect from these errors, we are implicitly looking at systems that are invariant under local transformations. Two objects that are equivalent under local transformations are said to be homologous, and the homology classes are the equivalence classes of objects up to local transformations.

For simplicity, we will focus on the homology classes of loops in a 2D torus (Fig. 5.3). For continuous spaces, local transformations are all continuous transformations. For curves, any continuous transformation corresponds to the boundary of a surface, namely the surface delimited by the curve before and the one after the transformation. This allows us to translate the same intuition to embedded lattices, where continuity is lost in a surface like the torus. In a lattice, curves are strings of edges and local transformations are the ones that delimit an area on the lattice. The homology classes we consider are thus the equivalence classes of closed curves up to deformations by boundaries of areas. There always exists a homology class of the empty curve that contains all boundaries of an area. In order to have more than one homology class, there must exist closed curves that are not boundaries of any area.

In a sphere, all closed curves are boundaries, and thus there is only one homology class. In a torus there exist two independent loops that are not boundaries of any surface. To enclose an area, two of these loops are needed, and they are the generators of four homology classes corresponding to the close curves equivalent to the empty curve, either of the two loops or of the union of them, as displayed in Fig. 5.3. We will distinguish the independent loops as the $\pm$ loops.

Topological Codes

Some topological error correction codes exploit the three concepts above by defining the logical bitflip and phaseflip operators as closed string operators around the generators of the homology classes. In practice, while the connectivity is local, connecting physical qubits in a toric topology is still hard, as the qubits will likely lie on a flat chip and the connectivity at the boundary will not be local anymore.

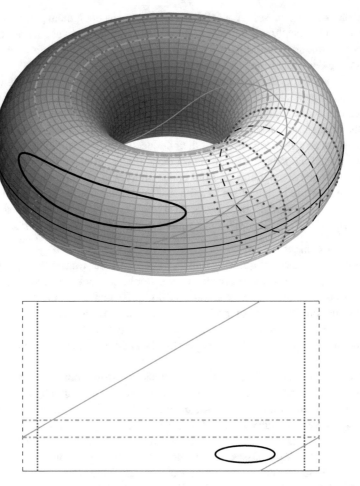

Fig. 5.3 Loops (closed strings) representing the homology classes of the torus embedded in 3D (top) and as a square with identified boundaries (bottom). Any closed string can be continuously deformed (is homologous) to one of the displayed loops. The difference between two deformations is always the boundary of a surface in the torus, like the black line, which is thus equivalent to the empty loop (they belong to the same homology class). A single green, purple, or orange loop is not the boundary of any surface, and they cannot be deformed into each other, but the orange loop is equivalent to the union of a green and a purple loop. An even number of green/purple loops is need to delimit a surface on the torus, and they form a basis for the homology classes on the torus. There are four homology classes corresponding to the parities of the numbers of green and purple loops. The same holds if we restrict the loops to strings of edges on a lattice embedded in the torus, like for a square or hexagonal lattice

For this reason, there exist variations of such codes on flat surfaces with a boundary that allow us to maintain the error correction properties while avoiding non-local connectivity on a chip [BK98, FM01]. Here, we only mention codes on a torus in order to more easily introduce the new concepts.

5.7.1   The Toric Code

The toric code [Kit97a, Kit97b, Kit03] is composed of qubits placed in a square lattice on a torus, with X and Z stabilizers computing the parities of four qubits of a unit square in a checkerboard pattern. For convenience, it is best to consider the square lattice where all qubits are placed on the edges rather than on the vertices. Each closed curve γ in this lattice defines a string of bitflip operators X^γ as the product of bitflips on each of the edge qubits of the curve. Each curve in the dual of this lattice similarly defines Z^γ. In the lattice, the dual curves look like curves where the edges are transversal to the curve, instead of parallel.

The plaquettes are the close curves around a square, and the stars are the crosses around a vertex (the closed curves around a dual square). The stabilizers of the toric code are all the bitflip parities X^p defined by plaquettes p and all the phaseflip parities Z^s defined by stars s, as shown in Fig. 5.4. The homology classes of closed curves in the lattice and dual lattice define four independent loops, $\pm$ for the lattice and $\pm$ for the dual, and thus four operators $X^\pm$ and $Z^\pm$ modulo plaquettes and star transformations. '+' loops in the two lattices always have an even number of intersections and similarly for '−' loops, while '+' and '−' loops always have an odd number of intersections. We thus have two pairs of anti-commuting (classes of) operators: $\{X^\pm, Z^\mp\} = 0$, and otherwise all operators commute with each other. The toric code thus defines two logical qubits.

All non-closed curves and dual curves are the errors. Since the logical operators are loops in the torus, the distance of the code is $\sqrt{n}$. However, the error rate does not decrease as we add more physical qubits, at best it will remain constant, namely it will scale with n. In principle, we would need an error correction code that can correct approximately n errors as we increase the number of qubits, and the exact value of the rate determines the necessary quality of the physical qubit implementation in order to successfully use the code. It is thus a priori not clear that the toric code is a good code for this task, but it is demonstrated to be a good code by the so-called threshold theorems [Kit97a, DKLP02], which show that it is still possible to send the logical error probability exponentially to zero, given that the error probability per physical qubit is below a certain threshold value. Namely, even though the distance does not scale linearly with the size of the code, errors are unstructured, and thus the probability of inducing a logical error is much smaller than the one implied by the code distance.

5.7.2   Color Codes

Color codes are topological stabilizer codes where qubits lie on the vertices of a *tricolorable* lattice [BMD06], like the hexagonal lattice in the torus of Fig. 5.5. In the hexagonal lattice, each hexagon cell defines both phaseflip and bitflip six-qubit stabilizer generators over the vertices of the cell.

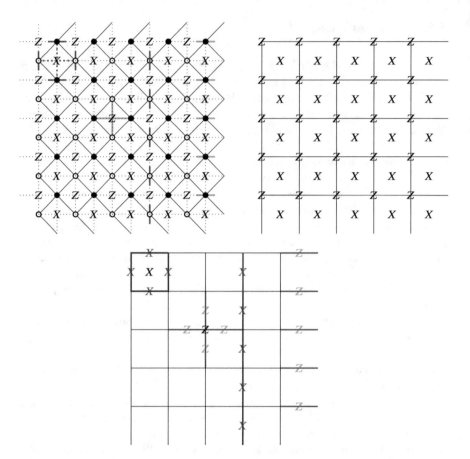

Fig. 5.4 The toric code (left), its stars/plaquettes depiction (right) and some basic string operators (bottom). *On the left*: the dots are physical qubits, and the X and Z represent ancilla qubits performing, respectively, X and Z parity measurements to the adjacent physical qubits connected with dotted lines. *On the right*: the physical qubits lie on the edges of the lattice, the Zs are called star operators and the Xs plaquettes operators. *At the bottom*: subset of edges represents tensor product of X (purple edges) and Z (green edges) operators; plaquettes (purple square) are boundaries of a square unit, stars (green cross) are boundaries of a dual square unit (plaquettes in the dual lattice); stars and plaquettes are the generators of the stabilizer group, which is the homology class of the empty string and empty dual string; the logical operators are the non-trivial loops of Xs (purple string) and dual loops of Zs (green string) in the torus

We use red (r), blue (b), or green (g) to tricolor the graph. Both the cells and the edges of the graph are *tricolorable*, with the color of the edge being opposite to the colors of the adjacent cells and edges connecting cells of the same color. We say that a curve is *colored* if all its edges have the same color. Colored curves are completely disconnected, and thus their boundaries coincide with their vertices. As such, denoting with ∂ the boundary operator on curves, which give the bitstring representing the vertices of the boundary, if γ is a colored curve, then $\partial\gamma$ gives

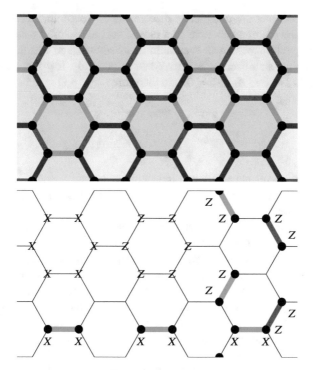

Fig. 5.5 Color code (top) and some of the stabilizer generators and color string operators (bottom). *At the top*: qubits lie on the vertices, the hexagonal lattice is tricolorable, and in particular the edges are tricolorable. Color strings are strings of edges of the same color, and they are completely disconnected; thus, their vertices are also their boundary. Each string of edges has three color strings: red, blue, and green completely disconnected substrings. *At the bottom*: two stabilizers and three color cycles (horizontal green Xs, vertical green Zs, and vertical red Zs) are shown. Each hexagon defines both a phaseflip and a bitflip six-qubit stabilizer generator as the parity over the six vertices (the boundary of a colored subset of the hexagon string of edges). All logical and stabilizer operators are color cycles (color substrings of cycles) of X or Z operators. Only two colors are independent, so it is enough to consider, e.g., red and green cycles. The logical operators are red/green X/Z operators over the two independent cycles of the torus for a total of eight logical operators. All logical operators of the same color overlap on an even number of qubits, and thus commute. Only X and Z logical operators of different color anti-commute, forming four logical qubits. In the picture, the horizontal green X and the vertical red Z form one of them

all the vertices touched by γ. Therefore, for any colored curve γ, their bitflip and phaseflip operators coincide with $X^{\partial\gamma}$ and $Z^{\partial\gamma}$. Because colored curves are completely disconnected, the boundary of a colored curve always has an even number of vertices and, in particular, if γ and ζ are colored curves of the same color, then they will overlap on an even number of vertices, and thus all their phaseflip and bitflip operators will commute: $[X^{\partial\gamma}, Z^{\partial\zeta}] = 0$.

We can partition any curve γ into the colored subcurves γ^b, γ^g, and γ^r composed of all the edges in γ of color b, g, and r, respectively. The following identity for the boundary of a partitioned curve holds $\partial\gamma^b \oplus \partial\gamma^g \oplus \partial\gamma^r = \partial\gamma$. Even though colored curves are completely disconnected, we say that a colored curve is closed if it is the colored subcurve of a closed curve. A closed colored curve will always overlap with faces on an even number of vertices and will therefore always commute with the stabilizer generators. Most importantly, if γ is a closed curve, it follows that $\partial\gamma^b \oplus \partial\gamma^g \oplus \partial\gamma^r = 0$, and therefore

$$X^{\partial\gamma^b} \cdot X^{\partial\gamma^g} \cdot X^{\partial\gamma^r} = \mathbb{1} \tag{5.22}$$

$$Z^{\partial\gamma^b} \cdot Z^{\partial\gamma^g} \cdot Z^{\partial\gamma^r} = \mathbb{1}. \tag{5.23}$$

Namely, closed curves have only two independent colors.[6]

We now choose red and blue as the independent colors. The combination of colored subcurves and homology classes of the torus create four different independent colored loop curves classes, namely $b_\pm$ and $r_\pm$. Therefore, we have doubled the number of homology classes.

As we have seen, operators over the same colored curves will always commute. Similarly, operators over the same loop, regardless of color, will overlap on an even number of qubits too and also commute. Only curves in $b_\pm$ intersect on an odd number of qubits with curves in $r_\mp$. This gives us the necessary commutation relation for the logical operators, and we thus have four logical qubits associated with the following four pair of operators:

$$\{X^{\partial b_\pm}, Z^{\partial r_\mp}\} = 0, \tag{5.24}$$

$$\{X^{\partial r_\pm}, Z^{\partial b_\mp}\} = 0. \tag{5.25}$$

In general, a hexagonal color code will encode double the number of logical qubits compared to the toric code analog (square lattice with stars and plaquettes) on the same surface but will have a lower distance and a lower threshold. The advantage of color codes is that they have easily constructible sets of transversal gates [KYP15], stemming from the fact that exchanging the role of phaseflip and bitflip stabilizers returns the same code, as opposed to the toric code where the exchange produces the toric code on the dual lattice. The code that alternates between a code and a subcode in order to achieve a universal set of transversal gates is indeed constructed using color codes [Bom15].

---

[6]The boundary of hexagon cells already has only two colors, which reduces the number of independent colors to a single one. This is why stabilizer generators are not associated with any color even though they are of the form $X^{\partial\gamma^c}$ and $Z^{\partial\gamma^c}$, where c is a color and γ is itself the boundary of a cell.

Chapter 6
Quantum Communication Networks: Design and Simulation

The theoretical design of the Internet and communication networks grew according to a layered architecture. Network tasks, goals, protocols, and functionalities were organized in a stack, which was divided into different layers according to their specific competences.

By thinking generally, a network is a set of interacting and interconnected elements (nodes), which performs a specific or some specific tasks, with particular behavior. The network's intrinsic interconnection/interaction implies the transmission of information among multiple nodes. Then, since communication of information is modeled as a shipping of commodities, additional problems such as setup of links for reliable transmission, access to communication links, and routing become something to be studied and analyzed. This implies the design of specific protocols and procedures for each task at each layer.

The complexity, variety, and intrinsic nature of quantum communication networks [Kim08, WEH18] are so different from the classical one that the usage of a known methodology may represent a limitation of the full potential of quantum mechanics. However, the mainstream current research on QCNs [WEH18, DSC+19] has maintained the same logic approach that has been applied to classical networks: a layered vision of the network functionalities based on a vertical protocol stack. This chapter, which considers the design of quantum networks, will mention this existing logic approach but will try to avoid a rigid classification into layers, in order not to force a specific vision onto the reader. In fact, the scope of this book is to give more general means, so that the reader is free to think about different potential architectures.

A network realizing quantum communications is a hybrid network. One possible implementation of the quantum network is using the quantum channels to distribute entangled quantum systems among the parties, so that the communication can be performed via teleportation. In order to do this, side by side, classical networks are employed to transmit the result of measurements in the teleportation protocol from

R. Bassoli et al., *Quantum Communication Networks*, Foundations in Signal
Processing, Communications and Networking 23,
https://doi.org/10.1007/978-3-030-62938-0_6

Fig. 6.1 Example of quantum communication between a source and a sink. The logic scheme also represents the different encoding techniques of classical bits and the respective Holevo bound χ

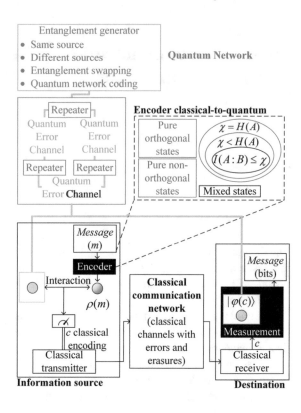

a source to a destination. The needed entangled quantum systems can be generated with direct interaction, or indirectly via entanglement swapping.

Figure 6.1 depicts a logic structure of a quantum communication based on first generation quantum repeaters and teleportation. The quantum network is mainly used to generate entangled quantum systems and to distribute them to the communicating nodes of the network. The distribution of entanglement happens via quantum links (which can experience quantum errors, as previously explained in Chap. 4) and quantum nodes, usually called quantum repeaters. These nodes are necessary since entanglement is reduced during propagation (i.e., quantum decoherence). The important metric for the quantum links and the quantum network in this setting is the entanglement generation rate.

Even if quantum systems are used to communicate, humans use classical representation of information, namely, classical bits. In fact, quantum states can be *prepared*, but any experiment/communication always starts with a classical choice, since we have no access to quantum states: a known quantum state as input cannot be assumed. The input is always a classical initial value. For the same reason, an experiment/communication always ends with a classical output because quantum states are not accessible except via measurements. Therefore, a general quantum task in the network from start to end will look as follows.

The information source needs an encoder to encode the classical message into qubits ($|\phi_1\rangle$). The encoding process can map the original bits into any quantum state (pure orthogonal, pure non-orthogonal, mixed quantum states, etc). Depending on the choice, the outgoing quantum state can carry a different amount of accessible information (i.e., the maximum mutual information between the classical input and an output measurement, upper bounded by the Holevo's bound χ). Next, the encoded quantum system $|\phi_1\rangle$ interacts with other systems (particle 2 in the figure) that are entangled with the quantum systems owned by the destination (particle 3). These entangled systems (particles 2 and 3) are the ones that were initially distributed via the quantum network. Particles 2 and 3 are normally in an EPR state, in order to maximize the upper bound χ.

The composite quantum system of particles 1 and 2 is measured at the source. This measurement affects the entangled quantum system of particle 3, so that, conditioned on the measurement outcome, the information is then *teleported* to the destination. However, entanglement is a local resource, so the destination cannot get any information from particle 3 only. At this stage, the classical communication becomes important. As explained in previous chapters, the measurement has a classical output. For each teleported qubit, two classical bits are sent to the destination via the classical transmitter. These classical bits have to go through classical links and nodes, and thus, performance metrics of communications via quantum teleportation are also affected by the classical metrics of classical channels and network nodes (e.g., errors and erasure). Finally, once the bits arrive at the classical receiver, they are used to perform the correct measurement on $|\varphi_3(c)\rangle$ in order to obtain the right information from the entangled qubit.

At first sight, this might seem exceedingly complicated, after all, why not simply send the classical data through the classical network? The reason is that the encoding might be hiding some quantum preprocessing of a prepared state, while the decoding might hide some quantum algorithm offered by a server.

As mentioned above, the distribution of entangled quantum systems among distant sources and destinations requires the deployment of quantum repeaters that are used to convert large numbers of short-distance and low-fidelity Bell pairs into smaller numbers of long-distance and high-fidelity pairs (see Chaps. 2 and 4). In fact, efficient quantum communication over long distances (≥ 1000 km) is still an open research issue due to fiber attenuation and operation errors accumulated over the entire communication distance. As an example, let us consider the distribution of entangled photons via optical fibers (wavelength $\lambda = 1550$ nm and attenuation $a = 0.2$ dB/km). The probability of successfully distributing entanglement can be expressed as

$$p_{ent} = 10^{-\frac{aL}{10}}, \tag{6.1}$$

where L is the distance. By considering entanglement distribution between two geographical points, with $L = 450$ km, the success probability becomes $p_{ent} = 10^{-9}$. The communication rate decreases as distance increases [YCL$^+$17, Cal17].

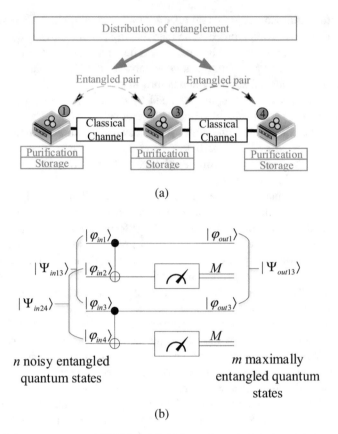

(a)

(b)

Fig. 6.2 (**a**) Quantum repeaters. (**b**) Bilateral CNOT transformation performed during purification

In the case of quantum communications via photons, this is due to the attenuation of the signal-to-noise ratio (SNR), and no amplification can be applied because of the no-cloning theorem. Thus, the most efficient solution that has been proposed by now is the division of a long link into shorter sublinks, interconnected via quantum repeaters.

Figure 6.2a represents an example of deployment of quantum repeaters. The network of repeaters is also interconnected via both quantum and classical channels. Currently, the goal is to place quantum repeaters every 10–100 km on quantum links and equip them with quantum memories realized by atomic or solid state qubits [vLAB⁺19]. The split of long point-to-point links can be realized through the deployment of terrestrial or satellite-based quantum repeaters. However, so far no working quantum repeater has been demonstrated, and they are thus an open research challenge.

6.1 Distillation in Quantum Repeaters

The general model of a first generation quantum repeater [VNM07, VLMN09, IG12, vLAB$^+$19] consists of two main procedures: *distillation* of high-fidelity EPR pairs from previously created noisy EPR pairs and teleportation between neighboring nodes. The input to the procedure of purification consists of n arbitrarily entangled states (e.g., Bell pairs), while the output is close to m maximally entangled states, with $m < n$ (i.e., higher-fidelity pairs). It can be considered a procedure of error correction.

Figure 6.2b depicts the quantum circuits for purification with the recurrence protocol [BDSW96]. This circuit represents a bilateral CNOT transformation and measurement, in which n noisy inputs become m maximally entangled outputs. If an error occurs during transmission of EPR states between the source and the sink, the output M of the measurement might differ. As shown in Fig. 6.2b, the output of a pure EPR state needs two input EPR states. The classical M bits that are the output of measurements are exchanged via classical channels in Fig. 6.2a.

The efficiency of purification is determined by two main factors: the choice of the quantum algorithm to be applied on each pair of Bell pairs, and the selection of the scheduling technique that significantly affects the number of required physical resources and the speed at which fidelity grows. Moreover, the quantum algorithm can be fixed or adaptive. The latter solution changes the algorithm according to the process of accumulation of noise in each pair in the repeater.

There are different scheduling algorithms for purification, which can be grouped into four main categories [IG12] (see Figs. 6.3 and 6.4):

- *Symmetric scheduling algorithm.* The name symmetric comes from the fact that the procedure starts at time t_1 with four Bell pairs of the same fidelity F_1. Next, at time t_2, the first two pairs get a successful purification, generating an EPR pair of fidelities $F_2 > F_1$. However, the successful outcome of purification is probabilistic, so the one applied on the second pair of Bell pairs can fail. Subsequently, the process takes two new Bell pairs with fidelity F_1 at times t_3 and t_4, and it again performs purification. If this time the output is successful, this result is purified with *Bell pair 5*, in order to obtain a Bell pair of greater fidelity ($F_3 > F_2$). Such a symmetric algorithm imposes, in practice, strict minimum hardware requirements and memory degradation, together with an inflexible allocation of resources.
- *Pumping scheduling algorithm.* The label pumping means that at each step of the algorithm and at each time t_i a fresh Bell pair is consumed to increase the fidelity of another fixed Bell pair. At each step, if the result is successful, a new Bell pair is used for an additional purification, in order to continuously increase the fidelity. This scheduling algorithm becomes especially suitable when fresh pairs have a relatively high fidelity and/or are affected by, for example, phase noise only.

Symmetric scheduling

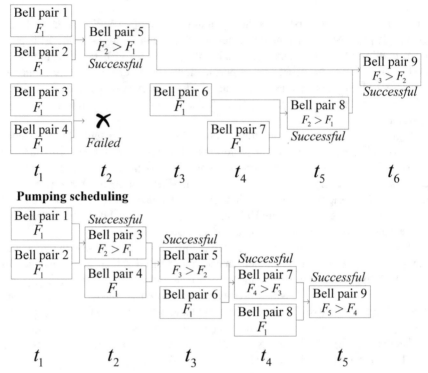

Fig. 6.3 Logic schemes of symmetric and pumping scheduling algorithms for purification. Variable F refers to the value of fidelity and t_i represents the time

- *Greedy scheduling algorithm.* The main objective of this algorithm is efficiency. In fact, each step uses previous successful results as input. In particular, it targets purification of all possible resources by achieving the highest value of output fidelity (F_5). An important aspect to be considered when using this approach to scheduling is that the fidelity of input EPR pairs should be high enough to avoid a greater number of failures, and so lower output fidelity. Greedy scheduling algorithms (together with pumping) can match Bell pairs with very different values of fidelity, reducing the success probability of purification and slightly increasing values of fidelity.
- *Banded scheduling algorithm.* Given the input Bell pairs, their fidelity space is divided into different subsets, called bands, and only Bell pairs within the same band are used for mutual purification. If at a specific time t_i there is no other EPR pair with the required fidelity, the procedure waits until a suitable EPR pair is obtained (see dashed lines in Fig. 6.4). At step t_5, the banding structure allows *Bell pair 9* to purify with *Bell pair 10*, whereas the symmetric algorithm would temporarily stop. Unlike pumping and greedy scheduling algorithms, banded

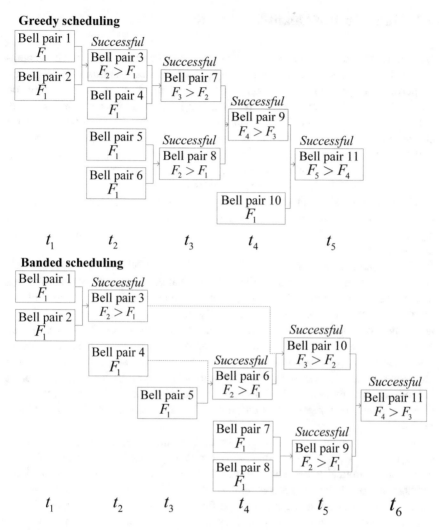

Fig. 6.4 Logic schemes of greedy and banded scheduling algorithms for purification. Variable F refers to the value of fidelity and t_i represents the time. The dashed lines in the banded scheduling algorithm mean the time Bell pairs have to wait for the creation of a suitable partner

scheduling assigns greater importance to Bell pairs with higher fidelity since they are more valuable than lower-fidelity Bell pairs. In fact, this algorithm only uses Bell pairs with higher fidelity when another Bell pair with similar value of fidelity is available. This helps improve the probability of the success of purification.

6.2   Taxonomy of Quantum Repeaters

Quantum repeaters proposed so far can be classified into three generations [MATN15, BvL19]. First, quantum repeaters where quantum error detection
is performed via sequential rounds of entanglement purification, acting on
copies of entangled states (concurrently using local quantum logic and classical
communications). Second, are quantum repeaters that apply quantum error
correction on errors in quantum memories. Third, we have quantum repeaters
without memories, where encoded quantum information is directly sent through
the channel (e.g., all-optical schemes). The majority of the existing practical
implementations follow the first kind of repeaters.

Unlike repeaters' architecture depicted in Fig. 6.2a, another specific and important family of quantum repeaters have been developed, which are called hybrid
repeaters. The reason for this name comes from using matter signals and light probes
and/or discrete and continuous quantum variables.

Hybrid quantum repeaters [vLLS$^+$06, LvN$^+$06, vLLMN08, ATL15, MLK$^+$16,
BvL19] use atomic-qubit entanglement and optical coherent state communication.
These repeaters are connected by optical fibers as quantum channels (called qubus),
which interact with the quantum state. Hybrid quantum repeaters provide the
advantages of discrete and continuous variable quantum states. Two-level atomic
systems with long coherence times are employed as quantum memories, while
optical coherent states are used to generate the initial entanglement between the
atoms using dispersive light-matter interactions and, in particular, highly efficient
homodyne measurements. This solution has the capability to be practically more
effective than quantum repeaters based on the generation and detection of single
photons.

The entanglement generated by hybrid quantum repeaters is an atom-light
entanglement, which employs quantum correlations between a discrete spin and
a continuous optical phase quadrature, instead of a discrete single photon. After
propagating to the closest repeater, an interaction with a second spin is realized, and
next, measurement on the light field is performed via homodyne detection. Finally,
entanglement between the two hybrid repeaters is obtained after the measurement
of the coherent state at the second repeater. The atomic qubits are stored in cavities.

The performance analysis of repeaters involves some fundamental metrics.
Figure 6.5 depicts the probability of success of the purification process for quantum
repeaters [MATN15] and hybrid quantum repeaters [IG12], respectively, while
also showing their trend for fidelity greater than 0.6. The number of rounds of
purification also determines the number of initial required Bell pairs. By setting
this value of available resources, the number of purification steps decreases the
entanglement rate.

Entanglement rate is another important metric to evaluate a network of quantum
repeaters for entanglement distribution. In particular, this value is inversely proportional to both the minimum time to generate an EPR state over distance between
repeaters and the average number of steps required to generate entanglement in all
2^n station pairs, given the value of success probability.

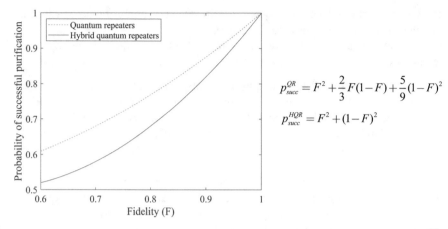

$$p_{succ}^{QR} = F^2 + \frac{2}{3}F(1-F) + \frac{5}{9}(1-F)^2$$

$$p_{succ}^{HQR} = F^2 + (1-F)^2$$

Fig. 6.5 Comparison between probability of successful purification of quantum repeaters p_{succ}^{QR} and of hybrid quantum repeaters p_{succ}^{HQR} [MATN15, IG12]

6.3 Storage in Quantum Repeaters

As previously shown in Fig. 6.2a, another fundamental task of repeaters is storage of quantum states. In classical systems in general, storage refers to a device (memory) that is used to save and to keep information (in the form of a number of bits) for a certain amount of time. In particular, the bit is a physical system that has two possible states, and it remains reliable as long as the distance between the two stable states is big enough to keep them distinct. In this sense, no noise or thermal agitation will be able to flip the bit state.

However, as defined in Sect. 2.1.2, quantum networks store, process, and communicate information via qubits. Because of their physical structure, qubits are very sensitive to interactions with external entities (i.e., any other simple composite quantum system), so that decoherence has to be avoided or significantly reduced via isolation or other physical means, in order to avoid noise. Important performance indicators are: coherence time (determined experimentally), memory cut-off time (i.e., maximally allowed storage time until quantum memory is reset and reinitialized, which affects memory dephasing errors on the entanglement fidelity), and maximal storage time (i.e., the time a quantum state is maintained in the quantum memory).

In general terms, *quantum memories* [IG12, DiV15, GI19] can be defined as n stationary quantum states, stored in quantum registers for information processing and retrieval. The main challenge in developing quantum memories is to provide reliable and durable storage/computation via unreliable components, since quantum memories will never be as reliable as classical ones. Such a problem has become solvable because it has been translated into the design of suitable error correction schemes (see Chap. 5) for reliable and fault-tolerant noisy quantum computation. In particular, the research community has identified the potentials of topological codes and the so-called color codes (see Sect. 5.7).

The most common metrics to evaluate the performance of quantum memories are as follows:

- *Conditional fidelity*. The definition of fidelity, used in the context of quantum memories, is the same one that quantum information theory uses (as described in Chap. 4). When talking about quantum memories, conditional fidelity means the overlap between the ingoing photon and the one that is recovered from the memory. The condition in this definition of fidelity is the successful re-emission of the photon.
- *Efficiency*. It is the probability of re-emitting a photon that was previously stored.
- *Storage time*. It is the time a quantum state can last stored in a quantum memory. It is lower bounded by the average entanglement creation time and the communication time between distant nodes.
- *Bandwidth*. This metric is responsible for the achievable repetition rate and the multiplexing potential of a quantum memory.
- *Capacity to store multiple photons and dimensionality*. It is the maximum number of photons (modes) that can be stored.
- *Wavelength*. This parameter is significantly important in long-range quantum communications. In fact, photons propagating over long distances in fibers are affected by absorption, so their wavelength should be in the region of small absorption. On the other hand, this does not happen to the same extent in free-space communications, for example, when memories are in satellites.

Quantum memories can be classified into different groups according to their physical implementation [LST09, SAA+10, IG12]. The following description lists and highlights the major technologies of quantum memories, with their principal characteristics.

Single-photon quantum memories are used in quantum repeaters based on the Duan–Lukin–Cirac–Zoller (DLCZ) procedure. This method consists of spontaneous Raman emission of a photon, creating a spin excitation in the atomic ensemble. Such a correlation between emitted photons and atomic excitations in each ensemble is the starting point for the generation of entanglement between distant ensembles, performed through single-photon detection.

Recent practical realizations [WLZ+19] show memory retrieval: both qubit fidelity in the range 0.99–0.88 and storage efficiency in the range 86%–68%, for storage time between 0.1 and 7 μs. In fact, a distinction between efficiency and fidelity should be specified for these kind of quantum memories. The former is the probability to emit a photon, while the latter identifies the "distance" between the emitted and original photons. In single-photon quantum memories, two possible failures can happen: failure in re-emitting a photon (erasure) or failure in the quality of the re-emitted photon, whose quantum state has an imperfect match with the original photon state that was stored. Experimental realization of 105-qubit memory was recently achieved in [JPC+19].

Another physical implementation of memories is via *rare-earth ions in solids* [TAC$^+$10], which exploit solid state atomic ensembles such as rare-earth-ion doped inorganic crystals or glasses. The important characteristics of these materials rely on the absence of motion of the absorbers and the excellent optical and hyperfine coherence times at cryogenic temperatures, which increase the storage time.

In order to control the negative effects of inhomogeneous dephasing, two methods have been proposed: controlled and reversible inhomogeneous broadening (CRIB) and atomic frequency comb (AFC). The former broadens an original narrow absorption line, which matches the spectral width of the light that is to be absorbed by exploiting both the so-called linear Stark effect and an applied external electric field gradient. Next, light is re-emitted and stimulated by a variation of the sign of the field, in order to invert the frequencies of atomic transitions around the central one. The latter adjusts the profile of absorption of an inhomogeneous and broadened solid state atomic medium through a series of periodic absorbing peaks. All the atoms in the comb absorb the photon to be stored so that the state of the light is transferred to collective atomic excitations at the optical transition. Such a periodic absorption profile implies a rephasing according to the comb spacing. Once all the atoms are again in phase, the photon is re-emitted using the collective interference between the emitters.

Experimentally speaking, memories based on rare-earth ions in solids can achieve conditional fidelity up to 95%, maximal efficiency of 45% for the CRIB, and 34% for the AFC (theoretically up to 100% is achievable in both cases), storage time of around 15 µs, bandwidths up to several hundreds of MHz, storage of large numbers of photons in the same memory, and wide range of wavelengths from 580 nm to 1530 nm. The read-out delay for a single stored photon can be chosen, and the existing implementations use standard lasers and fiber-optic components in a crystal, operating temperatures being in the region of 1–4 K.

Crystallographic defects in diamonds are impurity atoms in a diamond, allowing optical transitions with absorption wavelength from the UV to the infrared. These defects are deep within the bandgap of a diamond, which imply narrow and stable room temperature optical emission and excitation lines. Moreover, since such defects are electron paramagnetic, it becomes possible to exploit superior relaxation and decoherence properties of electron and nuclear spins in quantum memories. Lattices of diamonds are diamagnetic and rigid so that they can provide long electron and nuclear spin dephasing times (in the order of ms) at room temperatures. A well-known defect is the nitrogen vacancy (NV).

Calculating the fidelity of quantum memories based on crystallographic defects in diamonds that store single or multiple photons in defects is still an open challenge. The storage time is around 1–10 ms. Bandwidth is similar to that of rare-earth ions in solids, but in this case, the possibility of frequency tuning permits frequency multiplexing. The wavelength is around 700 nm; however, free-space quantum communication with NV defects works at 637 nm. Complexity is reasonable, while practical implementations are still at the beginning.

Semiconductor quantum dots can realize quantum memories by using the existing technologies based on semiconductors. This approach can be used to create fast and efficient quantum memories. During the storage of pure spin, fidelity can achieve values of around 0.80. The efficiency is determined by the input and output, coupling efficiency for the photon to dot-confined excitation. The storage time depends on the decoherence time of the electron spin stored in the quantum dot. This value can be greater than $100 \,\mu s$. Next, the bandwidth can reach 1 GHz; however, the operational speed of the gates controlling the device, hole tunneling times, and the excitation's radioactive lifetime can limit that value. Multiple-mode storage capacity is possible with multiple dots of different sizes; nevertheless, the storage of multiple photons in a single quantum dot scheme is hard to realize. The wavelength is in the near-IR, around 900 nm. Regarding complexity, the production of these memories is based on standard semiconductor fabrication techniques, but operations of the quantum memory happen at liquid-helium temperatures (i.e., less than 50 K).

Quantum memory can be based on *electromagnetically induced transparency* (EIT) [Luk03], which is a nonlinear optical effect in atoms that uses the strong interaction between light and atoms. Such interaction is a function of the frequency of light. Once the frequency matches the one for an atomic transition, a resonance happens and the optical response of the medium is enhanced. Light propagation is also accompanied by strong absorption and dispersion. The approach EIR makes an opaque medium transparent via quantum interference. In case of memories, a polariton propagates in an EIT medium, preserving its amplitude and shape. A control beam changes properties according to its intensity. The quantum state is mapped onto long-lived spin states of atoms. Next, the stored quantum state can be easily retrieved by reaccelerating the polariton.

An important kind of quantum memory exploits *atomic gases* [HSP10] such as room-temperature gas, ultra-cold gas, and Raman gas. The first is important since one of the main practical drawbacks of quantum memories is that they need very low temperatures. As an example, memories using alkali gases work at room temperature. The information is stored in the spin wave, which is not uniformly distributed. This implies that some atoms in a cell contribute more to the quantum memory than others. However, real implementations of this technical solution require many atoms (in the order of 10^{12}). On the other hand, solutions based on ultra-cold atoms need a much smaller number of atoms (in the order of 10^5). The high efficiency of such an approach relies on the very dense structure of cold atoms, which results in the necessity of fewer photons and shorter incoming light pulses. The third solution based on gases exploits Raman scattering. Such inelastic scattering can be employed to entangle quantum systems through light-matter coupling. This approach permits the usage of both room-temperature and ultra-cold atoms. This phenomenon is not only used for storage but can also be exploited to entangle remotely separated atoms.

In the above discussion, the complexity of the investigation of quantum storage and design of effective quantum memories is clear. Such a research problem has been addressed from different perspectives: considering theoretical physics, many-body/many-particle theory, quantum error correction, and exotic objects in quantum physics (e.g., Majorana particles). Moreover, the main practical challenge is the realization of quantum bit memory lasting for a long time at room temperature (nowadays, some solutions have been found exceeding a single-second lifetime).

6.4 Entanglement Distribution

Entanglement generation and distribution represent the first main steps of quantum communication based on teleportation. In fact, source and destination need to own entangled quantum systems in order to perform the subsequent steps to correctly finalize the communication of a qubit.

There are different ways to generate and to distribute entanglement. First, it is possible to generate entanglement of quantum systems (i.e., photons) that are generated by the same source. Experimentally, it can be made via a vacuum, which represents a generalization of beam splitting [AFP09].

Next, in the 1980s, experiments revealed that entanglement could also be generated among quantum systems (i.e., photons), which come from different sources. However, in the 1990s, the first experiment of entanglement swapping was shown, so that entanglement generation theory was further extended.

Entanglement swapping is a technique that allows us to entangle quantum systems that have never previously interacted. Figure 6.6 shows the procedure of entanglement swapping. The process starts with pump photons (photons of higher energy) passing through a nonlinear crystal that has an internal structure capable of splitting photon beams into pairs of photons. These entangled photon pairs (a signal and an idler photon) have lower energy. Such a procedure is called spontaneous

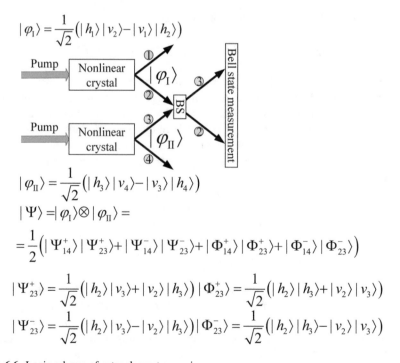

$$|\varphi_1\rangle = \frac{1}{\sqrt{2}}\left(|h_1\rangle|v_2\rangle - |v_1\rangle|h_2\rangle\right)$$

$$|\varphi_{II}\rangle = \frac{1}{\sqrt{2}}\left(|h_3\rangle|v_4\rangle - |v_3\rangle|h_4\rangle\right)$$

$$|\Psi\rangle = |\varphi_1\rangle \otimes |\varphi_{II}\rangle =$$

$$= \frac{1}{2}\left(|\Psi_{14}^+\rangle|\Psi_{23}^+\rangle + |\Psi_{14}^-\rangle|\Psi_{23}^-\rangle + |\Phi_{14}^+\rangle|\Phi_{23}^+\rangle + |\Phi_{14}^-\rangle|\Phi_{23}^-\rangle\right)$$

$$|\Psi_{23}^+\rangle = \frac{1}{\sqrt{2}}\left(|h_2\rangle|v_3\rangle + |v_2\rangle|h_3\rangle\right) \quad |\Phi_{23}^+\rangle = \frac{1}{\sqrt{2}}\left(|h_2\rangle|h_3\rangle + |v_2\rangle|v_3\rangle\right)$$

$$|\Psi_{23}^-\rangle = \frac{1}{\sqrt{2}}\left(|h_2\rangle|v_3\rangle - |v_2\rangle|h_3\rangle\right) \quad |\Phi_{23}^-\rangle = \frac{1}{\sqrt{2}}\left(|h_2\rangle|h_3\rangle - |v_2\rangle|v_3\rangle\right)$$

Fig. 6.6 Logic scheme of entanglement swapping

parametric downconversion (SPDC). The states of each photon pair are $|\psi_I\rangle$ and $|\psi_{II}\rangle$, respectively, while the state of the whole composite system, consisting of all the four particles, is $|\Psi\rangle$. The fact that quantum state $|\Psi\rangle$ is separable means that there is no entanglement between the photons of the EPR pairs $|\psi_I\rangle$ and $|\psi_{II}\rangle$.

Next, photons 2 and 3 pass through a suitable beam splitter (BS), and finally, a Bell-state measurement is performed, so that the state of particles 1 and 4 is projected onto a different entangled state, which is related to the result of measurement on particles 2 and 3: the four Bell states $|\Psi_{23}^+\rangle$, $|\Psi_{23}^-\rangle$, $|\Phi_{23}^+\rangle$, and $|\Phi_{23}^-\rangle$. By looking at the structure of $|\Psi\rangle$ in Fig. 6.6, it is possible to see that the projective Bell measurement on photons 2 and 3 projects photons 1 and 4 onto an entangled state with the same form.

Alongside this, a more recent way to generate and distribute entangled quantum systems is *quantum network coding* (QNC) [HIN+06, LOW10, SLPwL15, SINVM16, MSNVM18, LLY+19]. By considering more realistic quantum networks (i.e., with more complex topologies), the number of repeaters and links increases significantly as does cost of transmitting each qubit. Quantum network coding was proposed to reduce the costs of large-scale deployment of entanglement swapping. The employment of QNC can realize simultaneous quantum teleportation between far-spaced repeaters. In the case of a butterfly network, seven original Bell pairs permit teleportation between repeaters in opposite corners [SINVM16, MSNVM18].

Figures 6.7, 6.8, 6.9, and 6.10 show the various steps to distribute entanglement between two opposite quantum systems on a butterfly network.

The first step consists of the creation of a set

$$\{|\Psi_{1,2}\rangle, |\Psi_{3,4}\rangle, |\Psi_{5,6}\rangle, |\Psi_{7,8}\rangle, |\Psi_{9,10}\rangle, |\Psi_{11,12}\rangle, |\Psi_{13,14}\rangle\}$$

of Bell pairs at all nodes of the network, represented by the circuits composed of a Hadamard gate and a CNOT gate. Next, the procedure of connection is performed on three qubits, composed of a CNOT gate followed by measurement. The CNOT gate, respectively, *controls* qubits 3 and 7 via qubits 1 and 5. This means that the subsequent measurement on qubits 3 and 7 also affects the corresponding entangled qubits 4 and 8. The projective measurement operator depends on the normal Pauli operator X.

At this stage, parity creation is performed. Qubit 9 is first affected by the value of qubit 4 and then by the value of qubit 8 through a sequence of two CNOT gates. Following this, and considering that qubits 9 and 10 are entangled Bell pairs, the projective measurement and the associated normal Pauli operator X change the state of qubit 10. Such a qubit becomes the input control bit of two CNOT gates, which influence the values of qubits 11 and 13. The *addition* procedure ends with two projective measurements on the respective outputs of these CNOT gates, so that because of the original entanglement, qubits 12 and 14 are also modified.

Finally, the operation of fan-out and removal is performed via a sequence of Hadamard gates and projective measurement (depending on normal Pauli operator Z), so that two entangled Bell pairs are distributed to opposite repeaters in the network ($|\Psi_{16}\rangle$ and $|\Psi_{52}\rangle$).

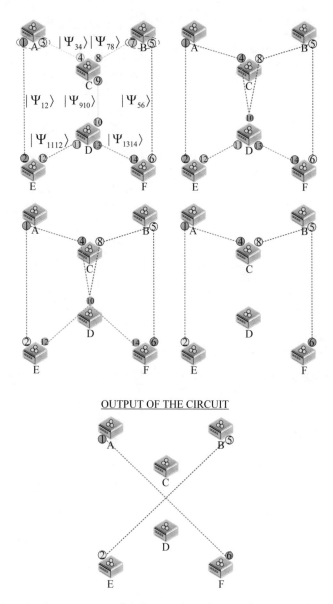

Fig. 6.7 Example of quantum network coding on a butterfly network

The characteristics of quantum network coding allow entanglement distribution to pass the bottleneck of link CD, which can be only solved via link multiplexing (e.g., time multiplexing) if entanglement swapping is employed. Moreover, QNC has an output of two Bell pairs from seven input Bell pairs. On the other hand, entanglement swapping among repeaters would have used six input Bell pairs.

Fig. 6.8 Example of quantum network coding on a butterfly network. Creation of Bell pairs and the connection stage

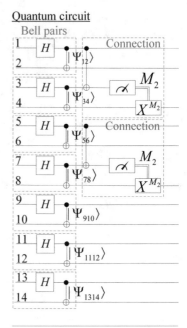

Fig. 6.9 Example of quantum network coding on a butterfly network. Parity creation and addition stages

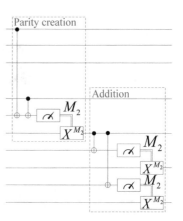

Given these premises, QNC outperforms entanglement swapping because it only needs one cycle of generation/distribution operations, while entanglement swapping requires two cycles to achieve the same outcome. Such a reduction in the number of cycles can become very important if the evaluation takes into account both the time to share entangled Bell pairs between neighboring receivers and the lifetime of entangled Bell pairs.

Fig. 6.10 Example of quantum network coding on a butterfly network. CNOT and FANOUT, remove and final output

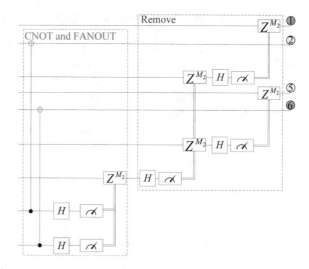

6.5 Multiple-Access Channel in Quantum Communication Networks

By now, the description and analysis of QCNs have been focusing on single-source unicast communications (see Fig. 6.1). However, recent works have started studying multi-source quantum communications, where more than one source node has access to (or *competes for*) the same transmission channel. Thus multiple-access channel techniques are also needed for quantum communication networks. In this context, both Nötzel and Leditzky et al. [Nö19, LALS20] investigated the case where two senders communicate with a receiver. It is important to notice that such research on MACs for two-source unicast quantum communications relies on a fundamental background, previously presented in Sect. 2.5 about CHSH inequality and quantum game theory (CHSH game).

Let us start by better clarifying the concept of the PR box [QS17], previously mentioned in Sect. 2.5. The PR box is a nonlocal box, also called a bipartite correlated box with two ends, owned by two quantum systems (Fig. 6.9). The relation between inputs and outputs can be expressed as

$$P_{PR}(a, b|x, y) = \begin{cases} 1/2, & \text{if } a \oplus b = xy \\ 0, & \text{otherwise.} \end{cases} \tag{6.2}$$

Such a box extends the concept of quantum nonlocality (*super-quantum correlation*), mentioned in Sect. 2.5, so that sharing a PR box allows System A and System B to always win the CHSH game. This is possible since the condition for the outputs a and b, produced by the PR box, is exactly the winning condition of the CHSH game. However, Sect. 2.5 showed that in reality, by using quantum entanglement, the success is reduced to $\mathcal{F}_q = \cos^2(\pi/8)$.

The CHSH value of the PR box is obtained by considering the two inputs/outputs of the box as the players' inputs/outputs of the CHSH game (see Fig. 2.12), while values in expression Eq. (2.123) are regarded as the expected value of the output of the PR box when System A and System B set their side of the PR box to 0 and 1. In fact, CHSH inequality defines the bound Eq. (2.123) on the statistics of spatially separated measurements by two experimenters on a physical state in LHV models. Such an information theoretic perspective of the scenario connects directly to the formalism of quantum measurement theory and quantum game theory.

Next, let us contextualize the concept of the PR box into the scenario of the MAC. Figure 6.10 depicts the information theoretic model of the MAC. Two sources require access to the common channel in order to send their information to the receiver. Relating to the PR box and the CHSH game, it means that sources need to win the nonlocal game Γ in order to maximize their achievable rates.

Nevertheless, the premises in Sect. 2.5 and the introduction above show that only a *super-quantum* strategy (PR box) can permit the achievement of the maximum sum rate $R_1 + R_2 = \log|\mathcal{X}_1| + \log|\mathcal{X}_2|$, where R_1 and R_2 are the respective rates of Source 1 and Source 2, and $\mathcal{X}_1$ and $\mathcal{X}_2$ are the sets of questions (from the game-theoretic point of view).

The solution of the MAC game, classically or quantum-mechanically, cannot allow the players (sources) to win the game but only to have success with a probability of 75 % and $\approx$ 85 % respectively. This implies a channel adding noise [LALS20] (or a jammer disturbing the communication [Nö19]) so that the achievable rates decrease (and the sum rate as well). In fact, the sum rate can be written as

$$R_1 + R_2 = I(X_1 Y_1 X_2 Y_2; Z) = H(Z) - p(\log|\mathcal{X}_1| + \log|\mathcal{X}_2|), \qquad (6.3)$$

where $X_1 Y_1$ is the input of Source 1, $X_2 Y_2$ is the input of Source 2, Z is the output at the receiver, and $p(\cdot)$ denotes the probability of losing the game Γ, given the product distribution on the questions and a strategy producing the answers. In particular, the value of $p(\cdot)$ can also be seen as the probability of failing the game $1 - \mathcal{F}$.

6.6 Classical Simulation of Quantum Communication Networks

After all the discussion in previous chapters, one realizes how much quantum-mechanical behaviors are different from classical ones, including in the telecommunication and computing fields. This means that even the design and setup of evaluation/emulation tools are not straightforward. Simulation and emulation, especially in computing and telecommunication, are fundamental tools for designing technologies for the future and for evaluating their performances in advanced, before actual development and deployment. This helps save costs and time especially. Nowadays, a universe of simulators and emulators are available for classical networks and computing simulations. Their software is very reliable, and most of the times is based on accurate mathematical models of the behavior of hardware and physics.

On the other hand, it is not possible to claim this for quantum mechanics and technologies based on this theory. At the moment, the realization of very basic components of quantum hardware is still an open research issue, and there are plenty of different non-standardized solutions. Moreover, at present, a full quantum network stack is missing, and no software development framework for writing quantum network applications exists.

Thus, the research community started implementing simulators of quantum communication networks, based on classical software. However, it is important to say: *classical reality is simulated by software on classical hardware so quantum reality should be simulated by specific software running on quantum hardware.* The curse of dimensionality will render extremely expensive to simulate large arbitrary quantum systems, and thus the simulation of large systems will either result in fundamental restrictions or large inaccuracies. The research focus then becomes evaluating such inaccuracies and being aware of classical assumptions and hypotheses behind the models that emulate quantum behaviors. The classical simulation of quantum communication networks is thus a field of research itself, as numerical analysis is a fundamental field for solving numerically mathematical problems. In such a context, a significant challenge is to provide an accurate *softwarization* of quantum systems and especially of quantum entanglement, which takes into account not only the algebraic viewpoint but also other perspectives (e.g., the one introduced by the application of tensor networks).

The following will review some of the existing well-known simulators for quantum communication networks and their principal characteristics. This list does not aim at being exhaustive but tries to provide the reader some knowledge about principal existing potential simulation platforms that can be used. Moreover, it has the scope of highlighting the existing issues and limitations (due to assumptions) that arise in the classical simulation of quantum communication networks. Some important frameworks such as *IBM Quantum simulators* [IBM19], *Microsoft Quantum Development Kit* (QDK) [Mic20], ProjectQ [Pro20] (from ETH Zurich), and Quantum Exact Simulation Toolkit (QuEST) [JBBB19] (from University of Oxford) are not described since they completely focus on quantum computing, without any mention of quantum communication networks.

6.6.1   SimulaQron

SimulaQron [DW17] is a simulator (freely available online) designed by QuTech at TU Delft. It is written in Python together with Twisted, which is a library providing functions to develop network applications in Python. Two libraries are used in SimulaQron via an interface, written both in Python and C#, called Classical

Quantum Combiner (CQC). This is a universal interface supporting *quantum* instructions such as commands to produce entanglement or transmit qubits. In this way, high-level applications can be implemented independent of the quantum hardware below.

SimulaQron considers the available direct communications between any two nodes in the network. In particular, it simulates quantum processors held by a source and receiver, communicating via a simulated quantum communication channel. Nevertheless, SimulaQron does not accurately model time-dependent noise in quantum communications. In order to simulate quantum entanglement, SimulaQron simulates the entire matrix ρ, while making two entangled qubits virtually available at two different nodes in the network. So, a network node has all qubits in the network in its own register, storing a matrix of dimension $2n \times 2n$ (n is the number of qubits).

SimulaQron enables platform-independent software development, thus without access to quantum hardware. SimulaQron emulates a quantum communication network by running each simulated quantum processor on a different classical real computer. Each virtual quantum network node locally processes quantum gates, measurements, and network-specific commands (e.g., as the previously mentioned generation of entanglement).

The main scope of SimulaQron is not an efficient simulation of a large-scale quantum communication network through its testing of quantum repeaters, quantum error-correcting codes, etc. The main aim is to provide to developers a high-level platform to write software that will be able to run on future eventual quantum hardware. SimulaQron does not easily synchronize the parties according to qubit arrival.

6.6.2 NetSquid

NetSquid [QuT17] (available online) is a collaboration between the Netherlands Organization for Applied Scientific Research (TNO) and the TU Delft. It is written in Python 3, using Numpy/SciPy, Cython, and C++. *NetSquid* is a *monolithic* simulator and does not provide a *realistic* simulation of quantum networks running on different *classical* computers. Moreover, it does not provide a real-time interactive experience.

It is a discrete-event simulator, focused on simulating delays in transmission, storage, and computing. Moreover, it includes modeling and evaluation of memory efficiency. The characteristic of being *discrete event* allows for modeling time so that it becomes possible to investigate the effect of noise in quantum communication networks. The main use cases are simulation of long quantum repeater chains and simulation of link-layer protocol for quantum networks (it is important to notice that no *standard* stack of quantum communication networks has been defined yet). NetSquid includes a network emulation mode, which can support high-level applications for quantum networks written in SimulaQron. That is possible via the CQC interface.

6.6.3   QuNetSim

QuNetSim [DNZB20] is a Python-based simulator for quantum networking, developed at the TU Munich. It is intended for developing and testing applications and protocols for quantum networks at network and application layers. This framework emulates quantum communications based and EPR generations over multi-hop quantum networks that may require potentially complicated routes. Network nodes are connected via modeled classical or quantum links. It also permits the application of synchronization methods on hosts. However, the current implementation of QuNetSim does not simulate the physical properties of quantum communication networks. In fact, it mainly simulates *the network layer and above*. At its current implementation stage, this simulator neglects most of the physical properties of quantum networks.

Regarding the assumption of a *layered network*, it is important to notice that there is no current counterpart of a classical ISO OSI stack in quantum communication networks and, especially, it is not known yet if such a communication model will be applicable. Nevertheless, QuNetSim is completely based on a network layering model inspired by the OSI model. Moreover, it implies the notion of packet, header, and payload. In fact, it models the control information of different payload types. This simulator assumes the existence of a *quantum packet* routed through the network, from source to receiver.

The structure of the simulator consists of network components for each layer, by logically separating distinct layers. Such components can be hosts, a transport layer for packet encoding and decoding, and the network itself. A host runs the applications, both classical and quantum, and thus it can process both kinds of information. Next, the transport layer object is devoted to creation of packets to be sent through the network. In this context, an assumption in the simulator's model takes into account that quantum information arrives with classical header information. Furthermore, quantum nodes can detect the arrival of qubits without destroying the related quantum information.

6.6.4   SQUANCH

The *Simulator for Quantum Networks and Channels* (SQUANCH) [Bar18] (open source) was developed at Stanford University. It is written in Python and NumPy. It simulates quantum communication networks with noisy quantum channels with configurable error models. Regarding functionalities, it is similar to SimulaQron, but it allows for setting parameters of the physical layer and error models.

SQUANCH simulates distributed quantum information processing that can be run in parallel for more efficient simulation. It simulates quantum communication networks with a specific focus on quantum transmissions and networking protocols. This simulator supports the emulation of many qubits, and it allows for customizable error models. Next, the quantum and classical network simulations are separate in order to permit modeling of a full hybrid classical-quantum infrastructure. A main difference between SQUANCH and QuNetSim is that the former cannot run sets of instructions in parallel.

The simulator is based on the concept of *separate agents*, handling subsets of a distributed quantum state in order to be more computationally efficient. In particular, agents are quantum network nodes, manipulating quantum information. An agent can include: a set of classical and quantum channels in order to reproduce interconnections with other agents, a classical memory, and a quantum memory. Next, it also has runtime logic for computing operations via gates and measurements.

Quantum gates are standard Python functions manipulating qubits as arguments. Then, ensembles are replicated by a *quantum stream* (QStream), storing the states of QSystems. A QSystem can represent the state of a multi-body, maximally entangled quantum system. It implements a qubits generator that can simulate qubit measurements in the computational basis. In parallel, a QStream (an iterable Python object) contains the number of qubits in each quantum system, and the number of systems in the stream. However, agents and channels are the classes that provide the abstraction of quantum nodes and connections.

The synchronization among communication entities presents some issues in SQUANCH. On the other hand, for example, QuNetSim provides each quantum transmitting node with an addressable quantum memory such that given an ID, a qubit can be recalled and manipulated.

6.6.5 SeQUeNCe

Simulator of Quantum Network Communication (SeQUeNCe) [Ket19] is a simulator of quantum networks developed by ARGONNE National Lab and the University of Chicago. It is a modular discrete-event simulator with a scheduler. It emulates quantum communications at the photon level. The discrete-event approach models the time-dependent operations of the system. In particular, each event is associated with a time and a function to be called.

This framework can model the characteristics of quantum hardware and network protocols, additionally emulating transmission of individual photon pulses and control messages with definition in the order of picoseconds. SeQUeNCe can also support polarization encoding of quantum information. This simulator was used to analyze the behavior of photonic quantum networks [WCK+19].

6.6.6 QuISP

Quantum Internet Simulation Package (QuISP) [Van20] (open source) is a simulator—that at the time of writing has just been released—created by the

Advancing Quantum Architecture (AQUA) research group at Keio University's Shonan Fujisawa Campus, Fujisawa, Japan. This software is based on an external C++ library, called Eigen, and OmNET++, from which it borrows all functionalities. QuISP can define complex network topologies and link parameters (e.g., length, channel error rates, numbers of qubit memories, gate error rates, etc.). Moreover, it provides a complete internal model of repeaters, including a connection manager, the RuleSet execution engine, real-time tracking of quantum states, etc.

The simulator emulates quantum procedures such as purification protocols (a single round of X purification, alternating X/Z purification, etc.) for single hops, and entanglement swapping. Moreover, it implements various generic networking-level aspects thanks to the OmNET++ baseline.

6.6.7   LIQUi|⟩

The *Language Integrated Quantum Operations* (LIQUi|⟩) [WS14] (freely available online) has been developed by the Quantum Architectures and Computation Group (QuArC) at Microsoft Research. This simulator targets algorithm writing and emulation of quantum hardware. In fact, it provides manipulation of quantum circuits, simulation, exporting, and visualization. It can also be used to research quantum noise, quantum error-correcting codes, circuit decomposition and optimization, classical control integration, and architecture-specific timing and layout constraints.

The software of LIQUi|⟩ uses F#, the F# interpreter, or C#, which can either compile it to run on a simulator or prepare it to be exported to other simulators.

LIQUi|⟩ has three main simulator modes. The first universal simulator allows the simulation of the quantum states and of any quantum circuit. In particular, it implements and makes readily available a large universal gate set. It is limited by the curse of dimensionality, and thus it can only simulate a small number of qubits. The second stabilizer simulator restricts the available gates to a generating set of the Clifford group, which can be simulated efficiently. It permits the simulation of large circuits on thousands of qubits. Since quantum error correction codes are included in this class of circuits, Stabilizer is the best simulator for the design and test of quantum error-correcting codes. The third physical simulator allows to implement various Hamiltonians and simulates the time evolution of the system by solving the equation of motion.

This simulator can be used for various use cases involving, for example, simple quantum teleportation, Shor's factoring, quantum chemistry (e.g., computing the ground state energy of a molecule), quantum error correction, quantum associative memory, and quantum linear algebra.

Chapter 7
Quantum Communication Networks: Final Considerations and Use Cases

As expressed in Chap. 1, quantum communication networks represent a break-through in telecommunications, which has acquired more and more interest from research communities of engineers and physicists. Their importance comes from the fact that quantum mechanics will upgrade existing and future classical commu-nication networks and computing devices via the exploitation of quantum resources such as qudits and entanglement.

With respect to the existing literature in the field, this book has presented quan-tum communication networks and their components in a new and comprehensive perspective, considering the role computing has had in classical future generation networks (e.g., 5G and beyond). The language and the notation have been accurately defined in order to simultaneously target readers with different backgrounds, such as natural sciences or information and communication technology.

As it was mentioned in Chap. 1, the applications of quantum communications in future communication networks are countless, and plenty of them are still unknown. In particular, Sect. 1.2 has displayed the need to actually deploy quantum mechanics in communication networks and distributed computing. Furthermore, it has also been underlined that the realization of a hybrid classical-quantum communication infrastructure will specifically help to figure out all the various research fields and businesses in communication and information technology that will be able to benefit from such a breakthrough.

In parallel, the research community has already proposed several use cases in which quantum communication networks could represent an actual powerful upgrade. The following will briefly describe them in order to give some starting points for readers who have understood the concepts and perspectives presented in all the previous chapters.

A first important use case of quantum communication networks can be *clock synchronization* [JADW00, Chu00, GLM01, dBB05, KKB+14, YCH18]. Why is this topic important? Accurate synchronization is the pillar of technologies such as

© The Editor(s) (if applicable) and The Author(s), under exclusive license to
Springer Nature Switzerland AG 2021
R. Bassoli et al., *Quantum Communication Networks*, Foundations in Signal
Processing, Communications and Networking 23,
https://doi.org/10.1007/978-3-030-62938-0_7

global positioning system (GPS), smart grid (synchronization of generators giving electric power to power grids), and telecommunications (synchronous communications, synchronization of base stations, FinTech, etc.). Especially when regarding synchronization in wireless cellular networks, new methods are needed to address the strict constraints of verticals in next-generation networks. Up to now, *classical* techniques that have been proposed are GPS-based retimers, embedded stand-alone clocks, and packet-based synchronization protocols. Furthermore, the outcomes of this topic of research will also be a pillar toward the realization of a more general paradigm in quantum distributed sensors: the Quantum Internet of Things (QIoT).

Clock synchronization means estimating the time difference between two clocks, placed at several spatially-separated entities. The accuracy of this procedure mainly depends on the stability of clock frequency and the delivery time of messages sent among different clocks. In this context, *classical* protocols can achieve accuracy lower than 100 ns. Nevertheless, these procedures can fail if uncertainty in the time for message delivery increases. In particular, clocks need $O(2^{2n})$ messages to obtain n digits of information of time difference.

By employing quantum communication networks, the spatially separated entities can share entangled qubits, which can be considered as a *quantum clock*. In fact, time information is maintained in the relative phase between the states. At the beginning of the procedure, only *entangled clocks* exist, which do not evolve in time. Next, information is extracted via measurements, and classical communications are performed by the separate nodes. Such procedures synchronize separate entities without exchanging time information.

On the other hand, a classical synchronization should have communicated actual time information over communication channels, whose imperfections would have limited the time accuracy. Furthermore, the synchronization via quantum communication networks makes the process independent of the uncertainties in time message exchanges. Thus, the communication complexity can decrease to $O(n)$ messages, therefore changing from exponential to linear. This does not require increasing any physical resources but only sharing entangled qubits. The key disadvantage of quantum clock synchronization is that the loss of a single qubit destroys the entanglement and renders the measurement useless. Additionally, repeated measurements and many qubits are necessary for the procedure of synchronization.

Another important application of quantum communication networks is the capability to beat classical ones in exchanging *Cartesian reference frames* among parties [PS01, RG03, BRS07]. Such *frames* represent spatial information, which is used to indicate a spatial direction. In the classical world, this information cannot be represented by a sequence of binary symbols, unless the source and the destination have prearranged information about a common coordinate system, specifying the values of relevant angles. By employing quantum communication systems, there are two main possibilities, either simply sending polarized spins along the direction to be communicated or (with higher accuracy) pre-distributing entangled quantum states.

Next, an application of quantum communication networks can be fundamental for the evolution of astronomy [Tsa11, GJC12]. Especially, interferometry among arrays of telescopes is a method to improve resolution, instead of using a single larger telescope. Nowadays, telescopes at different geographical points look at an astrophysical source that emits light incoherently, formed by a mixture of photons. This means that light collected by separate devices has to be physically combined by joining together all the optical paths.

In such a scenario, ground telescopes experience density fluctuations of the atmosphere, which modify the relative phase shift between the telescopes, thus completely saturating the signal. Moreover, such a physical combination is hard since sending photons over long distances produces losses of photons and uncontrolled phase shifts. On the other hand, if the array of telescopes exploits quantum communication networks, it becomes possible to use quantum repeaters to create entangled states shared among distant telescopes. The usage of entanglement as a resource can significantly improve interferometry at optical frequencies. Next, quantum information theory also plays an important role, since now telescopes classically process light. However, when the number of photons detected during observations is small, the quantum interpretation of light becomes fundamental.

References

[AG04]   Aaronson, S., & Gottesman, D. (2004). Improved simulation of stabilizer circuits. *Physical Review A, 70*(5), 052328.

[AvW18]   Abdou, A., van Oorschot, P. C., & Wan, T. (2018). Comparative analysis of control plane security of SDN and conventional networks. *IEEE Communications Surveys Tutorials, 20*(4), 3542–3559.

[ARS16]   Agiwal, M., Roy, A., & Saxena, N. (2016). Next generation 5G wireless networks: A comprehensive survey. *IEEE Communications Surveys Tutorials, 18*(3), 1617–1655.

[ABO08]   Aharonov, D., & Ben-Or, M. (2008). Fault-tolerant quantum computation with constant error rate. *SIAM Journal on Computing, 38*(4), 1207–1282.

[Ahl67]   Ahlswede, R. (1967). Certain results in coding theory for compound channels. In *Proceedings of the Colloquium on Information Theory* (vol. 1, pp. 35–60).

[Ahl78]   Ahlswede, R. (1978). Elimination of correlation in random codes for arbitrarily varying channels. *Zeitschrift für Wahrscheinlichkeitstheorie und verwandte Gebiete, 44*(2), 159–175.

[ABBN12]   Ahlswede, R., Bjelaković, I., Boche, H., & Nötzel, J. (2012). Quantum capacity under adversarial quantum noise: Arbitrarily varying quantum channels. *Communications in Mathematical Physics, 317*(1), 103–156.

[ABBN13]   Ahlswede, R., Bjelakovic, I., Boche, H., & Nötzel, J. (2013). Quantum capacity under adversarial quantum noise: Arbitrarily varying quantum channels. *Communications in Mathematical Physics, 317*(1), 103–156.

[AB07]   Ahlswede, R., & Blinovsky, V. (2007). Classical capacity of classical-quantum arbitrarily varying channels. *IEEE Transactions on Information Theory, 53*(2), 526–533.

[AD89a]   Ahlswede, R., & Dueck, G. (1989). Identification in the presence of feedback-a discovery of new capacity formulas. *IEEE Transactions on Information Theory, 35*(1), 30–36.

[AD89b]   Ahlswede, R., & Dueck, G. (1989). Identification via channels. *IEEE Transactions on Information Theory, 35*(1), 15–29.

[AW02]   Ahlswede, R., & Winter, A. (2002). Strong converse for identification via quantum channels. *IEEE Transactions on Information Theory, 48*(3), 569–579.

[AW69]   Ahlswede, R., & Wolfowitz, J. (1969). The structure of capacity functions for compound channels. In *Probability and Information Theory* (pp. 12–54). Berlin: Springer.

© The Editor(s) (if applicable) and The Author(s), under exclusive license to
Springer Nature Switzerland AG 2021
R. Bassoli et al., *Quantum Communication Networks*, Foundations in Signal
Processing, Communications and Networking 23,
https://doi.org/10.1007/978-3-030-62938-0

[AZ95]   Ahlswede, R., & Zhang, Z. (1995). New directions in the theory of identification via channels. *IEEE Transactions on Information Theory, 41*(4), 1040–1050.

[ANYG15]   Ahmad, I., Namal, S., Ylianttila, M., & Gurtov, A. (2015). Security in software defined networks: A survey. *IEEE Communications Surveys Tutorials, 17*(4), 2317–2346.

[ADJ17]   Anshu, A., Devabathini, V. K., & Jain, R. (2017). Quantum communication using coherent rejection sampling. *Physical Review Letters, 119*(12), 120506.

[AJW19]   Anshu, A., Jain, R., & Warsi, N. A. (2019). Building blocks for communication over noisy quantum networks. *IEEE Transactions on Information Theory, 65*(2), 1287–1306.

[AFP09]   Auletta, G., Fortunato, M., & Parisi, G. (2009). *Quantum mechanics* (1st ed.). Cambridge: Cambridge University Press.

[ATL15]   Azuma, K., Tamaki, K., & Lo, H. (2015). All-photonic quantum repeaters. *Nature Communications, 6*, 6787.

[BA12]   Band, Y., & Avishai, Y. (2012). *Quantum Mechanics with Applications to Nanotechnology and Information Science* (1st ed.). Cambridge: Academic.

[BBC$^+$95] Barenco, A., Bennett, C. H., Cleve, R., DiVincenzo, D. P., Margolus, N., Shor, P., et al. (1995). Elementary gates for quantum computation. *Physical Review A, 52*(5), 3457–3467.

[Bar18]   Bartlett, B. (2018). A distributed simulation framework for quantum networks and channels. arXiv:quant-ph/1808.07047.

[BRS07]   Bartlett, S. D., Rudolph, T., & Spekkens, R. W. (2007). Reference frames, superselection rules, and quantum information. *Reviews of Modern Physics, 79*(2), 555–609.

[BGADR19]   Bassoli, R., Granelli, F., Arzo, S. T., & Di Renzo, M. (2019). Toward 5G cloud radio access network: An energy and latency perspective. Transactions on Emerging Telecommunications Technologies e3669.

[Bel06]   Bellac, M. L. (2006). *Quantum physics* (1st ed.). Cambridge: Cambridge University Press.

[Ben73]   Bennett, C. H. (1973). Logical reversibility of computation. *IBM Journal of Research and Development, 17*(6), 525–532.

[BBBV97]   Bennett, C. H., Bernstein, E., Brassard, G., & Vazirani, U. (1997). Strengths and weaknesses of quantum computing. *SIAM Journal on Computing, 26*(5), 1510–1523.

[BB84]   Bennett, C. H., & Brassard, G. (1984). An update on quantum cryptography. In *Workshop on the Theory and Application of Cryptographic Techniques* (pp. 475–480).

[BBC$^+$93] Bennett, C. H., Brassard, G., Crépeau, C., Jozsa, R., Peres, A., & Wootters, W. K. (1993). Teleporting an unknown quantum state via dual classical and Einstein-Podolsky-Rosen channels. *Physical Review Letters, 70*(13), 1895.

[BDF$^+$99] Bennett, C. H., DiVincenzo, D. P., Fuchs, C. A., Mor, T., Rains, E., Shor, P. W., et al. (1999). Quantum nonlocality without entanglement. *Physical Review A, 59*(2), 1070–1091.

[BDSW96]   Bennett, C. H., DiVincenzo, D. P., Smolin, J. A., & Wootters, W. K. (1996). Mixed-state entanglement and quantum error correction. *Physical Review A, 54*(5), 3824–3851.

[BSST99]   Bennett, C. H., Shor, P. W., Smolin, J. A., & Thapliyal, A. V. (1999). Entanglement-assisted classical capacity of noisy quantum channels. *Physical Review Letters, 83*(15), 3081–3084.

[BvL19]   Bergmann, M., & van Loock, P. (2019). Hybrid quantum repeater for qudits. *Physical Review A, 99*(3), 032349.

[Ber10]   Bernstein, D. J. (2010). Grover vs. mceliece. In N. Sendrier (Ed.), *Post-quantum cryptography* (pp. 73–80). Berlin: Springer.

[BGW17]   Berta, M., Gharibyan, H., & Walter, M. (2017). Entanglement-assisted capacities of compound quantum channels. *IEEE Transactions on Information Theory, 63*(5), 3306–3321.

[BBJ19]   Bhattacharya, S., Budkuley, A. J., & Jaggi, S. (2019). Shared randomness in arbitrarily varying channels. In *2019 IEEE International Symposium on Information Theory (ISIT)* (pp. 627–631).

[BBJN13] Bjelaković, I., Boche, H., Janßen, G., & Nötzel, J. (2013). Arbitrarily varying and compound classical-quantum channels and a note on quantum zero-error capacities. In *Information Theory, Combinatorics, and Search Theory* (pp. 247–283). Berlin: Springer.

[BBN09]   Bjelaković, I., Boche, H., & Nötzel, J. (2009). Entanglement transmission and generation under channel uncertainty: Universal quantum channel coding. *Communications in Mathematical Physics, 292*(1), 55–97.

[BSS12]   Bjelakovic, I., & Siegmund-Schultze, R. (2012). Quantum Stein's lemma revisited, inequalities for quantum entropies, and a concavity theorem of Lieb. arXiv:quant-ph/0307170v2.

[BGS13]   Bocharov, A., Gurevich, Y., & Svore, K. M. (2013). Efficient decomposition of single-qubit gates into v basis circuits. *Physical Review A, 88*(1), 012313.

[BCCD14a] Boche, H., Cai, M., Cai, N., & Deppe, C. (2014). Secrecy capacities of compound quantum wiretap channels and applications. *Physical Review A, 89*(5), 052320.

[BCDN16] Boche, H., Cai, M., Deppe, C., & Nötzel, J. (2016). Classical-quantum arbitrarily varying wiretap channel: Common randomness assisted code and continuity. *Quantum Information Processing, 16*(1), 35.

[BCND19] Boche, H., Cai, M., Nötzel, J., & Deppe, C. (2019). Secret message transmission over quantum channels under adversarial quantum noise: Secrecy capacity and super-activation. *Journal of Mathematical Physics, 60*(6), 062202.

[BDNW18] Boche, H., Deppe, C., Nötzel, J., & Winter, A. (2018). Fully quantum arbitrarily varying channels: Random coding capacity and capacity dichotomy. In *2018 IEEE International Symposium on Information Theory (ISIT)*.

[BDW19]   Boche, H., Deppe, C., & Winter, A. (2019). Secure and robust identification via classical-quantum channels. *IEEE Transactions on Information Theory, 65*(10), 6734–6749.

[BJK17]   Boche, H., Janßen, G., & Kaltenstadler, S. (2017). Entanglement-assisted classical capacities of compound and arbitrarily varying quantum channels. *Quantum Information Processing, 16*(4), 88.

[BJS19a] Boche, H., Janßen, G., & Saeedinaeeni, S. (2019). Simultaneous transmission of classical and quantum information under channel uncertainty and jamming attacks. *Journal of Mathematical Physics, 60*(2), 022204.

[BJS19b] Boche, H., Janßen, G., & Saeedinaeeni, S. (2019). Universal superposition codes: Capacity regions of compound quantum broadcast channel with confidential messages. arXiv:quantph/1911.07753.

[BN14]   Boche, H., & Nötzel, J. (2014). Positivity, discontinuity, finite resources, and nonzero error for arbitrarily varying quantum channels. *Journal of Mathematical Physics, 55*(12), 122201.

[BSP18]   Boche, H., Schaefer, R. F., & Poor, H. V. (2018). Analytical properties of shannon's capacity of arbitrarily varying channels under list decoding: Super-additivity and discontinuity behavior. *Problems of Information Transmission, 54*(3), 199–228.

[Bom15]   Bombin, H. (2015). Gauge color codes: Optimal transversal gates and gauge fixing in topological stabilizer codes. *New Journal of Physics, 17*(8), 083002.

[BMD06]   Bombin, H., & Martin-Delgado, M. A. (2006). Topological quantum distillation. *Physical Review Letters, 97*(18), 180501.

[BAB+18]   Botsinis, P., Alanis, D., Babar, Z., Nguyen, H. V., Chandra, D., Ng, S. X., &
            Hanzo, L. (2018). Quantum search algorithms for wireless communications. *IEEE
            Communications Surveys & Tutorials, 21*(2), 1209–1242.
[BMP+00]   Boykin, P. O., Mor, T., Pulver, M., Roychowdhury, V., & Vatan, F. (2000). A new
            universal and fault-tolerant quantum basis. *Information Processing Letters, 75*(3),
            101–107.
[BKL+19]   Brandão, F. G. S. L., Kalev, A., Li, T., Lin, C. Y.-Y., Svore, K. M., & Wu, X. (2019).
            Quantum SDP solvers: Large speed-ups, optimality, and applications to quantum
            learning. In C. Baier, I. Chatzigiannakis, P. Flocchini, & S. Leonardi (Eds.), *46th
            International Colloquium on Automata, Languages, and Programming (ICALP
            2019). Leibniz International Proceedings in Informatics (LIPIcs)* (vol. 132, pp.
            27:1–27:14). Wadern: Schloss Dagstuhl–Leibniz-Zentrum fuer Informatik.
[BS17]   Brandao, F. G. S. L., & Svore, K. M. (2017). Quantum speed-ups for solving
            semidefinite programs. In *2017 IEEE 58th Annual Symposium on Foundations of
            Computer Science (FOCS)* (pp. 415–426).
[BK98]   Bravyi, S. B., & Kitaev, A. Y. (1988). Quantum codes on a lattice with boundary.
            arXiv:quant-ph/9811052.
[BCP+14]   Brunner, N., Cavalcanti, D., Pironio, S., Scarani, V., & Wehner, S. (2014). Bell
            nonlocality. *Reviews of Modern Physics, 86*(2), 419–478.
[BCMdW10]   Buhrman, H., Cleve, R., Massar, S., & de Wolf, R. (2010). Nonlocality and
            communication complexity. *Reviews of Modern Physics, 82*(1), 665–698.
[BIK+16]   Buzzi, S., Chih-Lin, I, Klein, T. E., Poor, H. V., Yang, C., & Zappone, A. (2016). A
            survey of energy-efficient techniques for 5G networks and challenges ahead. *IEEE
            Journal on Selected Areas in Communications, 34*(4), 697–709.
[HB19]   Cai, N., Boche, H., & Cai, M. (2019). Message transmission over classical
            quantum channels with a jammer with side information, correlation as resource
            and common randomness generating. In *2019 IEEE International Symposium on
            Information Theory*. Piscataway: IEEE.
[HB20b]   Cai, N., Boche, H., & Cai, M. (2020). Message transmission over classical
            quantum channels with a jammer with side information: Correlation as resource,
            common randomness generation. *Journal of Mathematical Physics, 61*, 062201.
[CWY04]   Cai, N., Winter, A., & Yeung, R. W. (2004). Quantum privacy and quantum wiretap
            channels. *Problems of Information Transmission, 40*(4), 318–336.
[Cal17]   Caleffi, M. (2017). Optimal routing for quantum networks. *IEEE Access, 5*,
            22299–22312.
[CKR09]   Christandl, M., König, R., & Renner, R. (2009). Postselection technique for
            quantum channels with applications to quantum cryptography. *Physical Review
            Letters, 102*(2), 020504.
[Chu00]   Chuang, I. L. (2000). Quantum algorithm for distributed clock synchronization.
            Physical Review Letters, 85(9), 2006–2009.
[CHTW04]   Cleve, R., Hoyer, P., Toner, B., & Watrous, J. (2004). Consequences and limits of
            nonlocal strategies. In *19th IEEE Annual Conference on Computational Complex-
            ity Proceedings, 2004* (pp. 236–249).
[CW90]   Coppersmith, D., & Winograd, S. (1990). Matrix multiplication via arithmetic
            progressions. *Journal of Symbolic Computation, 9*(3), 251–280. Computational
            algebraic complexity editorial.
[CK78]   Csiszár, I., & Körner, J. (1978). Broadcast channels with confidential messages.
            IEEE Transactions on Information Theory, 24(3), 339–348.
[CK81]   Csiszar, E., & Körner, J. (1981). *Information Theory: Coding Theorems for
            Discrete Memoryless Systems*. New York: Academic Press.
[DSC+19]   Dahlberg, A., Skrzypczyk, M., Coopmans, T., Wubben, L., Rozpędek, F., Pompili,
            M., et al. (2019). A link layer protocol for quantum networks. In *Proceedings of
            the ACM Special Interest Group on Data Communication, SIGCOMM '19* (pp.
            159–173). New York: Association for Computing Machinery.

[DW17]   Dahlberg, A., & Wehner, S. (2017). SimulaQron - a simulator for developing quantum internet software. arXiv:quant-ph/1712.08032.

[DCA+17]   Dargahi, T., Caponi, A., Ambrosin, M., Bianchi, G., & Conti, M. (2017). A survey on the security of stateful SDN data planes. *IEEE Communications Surveys Tutorials, 19*(3), 1701–1725.

[DL70]   Davies, E. B., & Lewis, J. T. (1970). An operational approach to quantum probability. *Communications in Mathematical Physics, 17*(3), 239–260.

[DWF16]   Dayarathna, M., Wen, Y., & Fan, R. (2016). Data center energy consumption modeling: A survey. *IEEE Communications Surveys Tutorials, 18*(1), 732–794.

[dBB05]   de Burgh, M., & Bartlett, S. D. (2005). Quantum methods for clock synchronization: Beating the standard quantum limit without entanglement. *Physical Review A, 72*(4).

[DKLP02]   Dennis, E., Kitaev, A. Y., Landahl, A., & Preskill, J. (2002). Topological quantum memory. *Journal of Mathematical Physics, 43*(9), 4452–4505.

[Dev05]   Devetak, I. (2005). The private classical capacity and quantum capacity of a quantum channel. *IEEE Transactions on Information Theory, 51*(1), 44–55.

[DS05]   Devetak, I., & Shor, P. W. (2005). The capacity of a quantum channel for simultaneous transmission of classical and quantum information. *Communications in Mathematical Physics, 256*(2), 287–303.

[DW05]   Devetak, I., & Winter, A. (2005). Distillation of secret key and entanglement from quantum states. *Proceedings of the Royal Society A: Mathematical, Physical and Engineering Sciences, 461*(2053), 207–235.

[DJLS13]   Dey, B. K., Jaggi, S., Langberg, M., & Sarwate, A. D. (2013). Upper bounds on the capacity of binary channels with causal adversaries. *IEEE Transactions on Information Theory, 59*(6), 3753–3763.

[DNZB20]   DiAdamo, S., Nözel, J., Zanger, B., & Mert Beşe, M. (2020). QuNetSim: A software framework for quantum networks.

[DiV00]   DiVincenzo, D. P. (2000). The physical implementation of quantum computation. *Fortschritte der Physik, 48*(9–11), 771–783.

[DiV15]   DiVincenzo, D. P. (2015). The memory problem of quantum information processing. *Proceedings of the IEEE, 103*(8), 1417–1425.

[EK09]   Eastin, B., & Knill, E. (2009). Restrictions on transversal encoded quantum gate sets. *Physical Review Letters, 102*(11), 110502.

[EHKS14]   Eisenträger, K., Hallgren, S., Kitaev, A. Y., & Song, F. (2014). A quantum algorithm for computing the unit group of an arbitrary degree number field. In *Proceedings of the Forty-Sixth Annual ACM Symposium on Theory of Computing, STOC'14* (pp. 293–302). New York: Association for Computing Machinery.

[EW00]   Eisert, J., & Wilkens, M. (2000). Quantum games. *Journal of Modern Optics, 47*(14–15), 2543–2556.

[EWL99]   Eisert, J., Wilkens, M., & Lewenstein, M. (1999). Quantum games and quantum strategies. *Physical Review Letters, 83*, 3077–3080

[Eri85]   Ericson, T. (1985). Exponential error bounds for random codes in the arbitrarily varying channel. *IEEE Transactions on Information Theory, 31*(1), 42–48.

[Fan73]   Fannes, M. (1973). A continuity property of the entropy density for spin lattice systems. *Communications in Mathematical Physics, 31*(4), 291–294.

[FTKS19]   Farris, I., Taleb, T., Khettab, Y., & Song, J. (2019). A survey on emerging SDN and NFV security mechanisms for IoT systems. *IEEE Communications Surveys Tutorials, 21*(1), 812–837.

[Fey99]   Feynman, R. P. (1999). Simulating physics with computers. *International Journal of Theoretical Physics, 21*(6–7), 467–488.

[FGS20]   Fitzek, F. H. P., Granelli, F., & Seeling, P. (2020). *Computing in communication networks - from theory to practice* (vol. 1., 1st ed.). Amsterdam: Elsevier. https://cn.ifn.et.tu-dresden.de/compcombook/

[FSG09]   Fowler, A. G., Stephens, A. M., & Groszkowski, P. (2009). High-threshold universal quantum computation on the surface code. *Physical Review A, 80*(5), 052312.

[FM01]   Freedman, M. H., & Meyer, D. A. (2001). Projective plane and planar quantum codes. *Foundations of Computational Mathematics, 1*(3), 325–332.

[FC72]   Freedman, S. J., & Clauser, J. F. (1972). Experimental test of local hidden-variable theories. *Physical Review Letters, 28*, 938–941.

[FVDG99]   Fuchs, C. A., & Van De Graaf, J. (1999). Cryptographic distinguishability measures for quantum-mechanical states. *IEEE Transactions on Information Theory, 45*(4), 1216–1227.

[GLM01]   Giovannetti, V., Lloyd, S., & Maccone, L. (2001). Quantum-enhanced positioning and clock synchronization. *Nature, 412*(6845), 417–419.

[GVW⁺15]   Giustina, M., Versteegh, M. A. M., Wengerowsky, S., Handsteiner, J., Hochrainer, A., Phelan, K., et al. (2015). Significant-loophole-free test of Bell's theorem with entangled photons. *Physical Review Letters, 115*(25), 250401.

[Got97]   Gottesman, D. (1997). *Stabilizer codes and quantum error correction.* Ph.D. Thesis, California Institute of Technology.

[Got98a]   Gottesman, D. (1998). The Heisenberg representation of quantum computers. In *Group22: Proceedings of the XXII International Colloquium on Group Theoretic Methods in Physics* (pp. 32–43). Cambridge: International Press.

[Got98b]   Gottesman, D. (1998). Theory of fault-tolerant quantum computation. *Physical Review A, 57*, 127–137.

[GJC12]   Gottesman, D., Jennewein, T., & Croke, S. (2012). Longer-baseline telescopes using quantum repeaters. *Physical Review Letters, 109*(7), 070503.

[GB18]   Granelli, F., & Bassoli, R. (2018). Autonomic mobile virtual network operators for future generation networks. *IEEE Network, 32*(5), 76–84.

[GW07]   Gutoski, G., & Watrous, J. (2007). Toward a general theory of quantum games. *Proceedings of the Thirty-Ninth Annual ACM Symposium on Theory of Computing - STOC'07.*

[GI19]   Gyongyosi, L., & Imre, S. (2019). A survey on quantum computing technology. *Computer Science Review, 31*, 51–71.

[Hal07]   Hallgren, S. (2007). Polynomial-time quantum algorithms for Pell's equation and the principal ideal problem. *Journal of ACM, 54*(1), 4.

[HSP10]   Hammerer, K., Sørensen, A. S., & Polzik, E. S. (2010). Quantum interface between light and atomic ensembles. *Reviews of Modern Physics, 82*(2), 1041–1093.

[HIN⁺06]   Hayashi, M., Iwama, K., Nishimura, H., Raymond, R., & Yamashita, S. (2006). Quantum network coding. In *STACS: Annual Symposium on Theoretical Aspects of Computer Science. Lecture Notes in Computer Science* (pp. 610–621). Berlin: Springer.

[HW12]   Hayden, P., & Winter, A. (2012). Weak decoupling duality and quantum identification. *IEEE Transactions on Information Theory, 58*(7), 4914–4929.

[Hel69]   Helstrom, C. W. (1969). Quantum detection and estimation theory. *Journal of Statistical Physics, 1*(2), 231–252.

[HBD⁺15]   Hensen, B., Bernien, H., Dréau, A. E., Reiserer, A., Kalb, N., Blok, M. S., et al. (2015). Loophole-free bell inequality violation using electron spins separated by 1.3 kilometres. *Nature, 526*(7575), 682–686.

[Hol73]   Holevo, A. S. (1973). Bounds for the quantity of information transmitted by a quantum communication channel. *Problemy Peredachi Informatsii, 9*(3), 3–11.

[Hol98]   Holevo, A. S. (1998). The capacity of the quantum channel with general signal states. *IEEE Transactions on Information Theory, 44*(1), 269–273.

[HHHH09]   Horodecki, R., Horodecki, P., Horodecki, M., & Horodecki, K. (2009). Quantum entanglement. *Reviews of Modern Physics, 81*(2), 865–942.

[HW09]   Hsieh, M.-H., & Wilde, M. M. (2009). Public and private communication with a quantum channel and a secret key. *Physical Review A, 80*(2), 022306.

[HW10a]   Hsieh, M.-H., & Wilde, M. M. (2010). Entanglement-assisted communication of classical and quantum information. *IEEE Transactions on Information Theory, 56*(9), 4682–4704.

[HW10b]   Hsieh, M.-H., & Wilde, M. M. (2010). Trading classical communication, quantum communication, and entanglement in quantum shannon theory. *IEEE Transactions on Information Theory, 56*(9), 4705–4730.

[IBM19]   IBM quantum experience (2019). https://quantum-computing.ibm.com/

[Ili89]   Iliopoulos, C. S. (1989). Worst-case complexity bounds on algorithms for computing the canonical structure of infinite abelian groups and solving systems of linear diophantine equations. *SIAM Journal on Computing, 18*(4), 670–678.

[IG12]   Imre, S., & Gyongyosi, L. (2012). *Advanced Quantum Communications: An Engineering Approach* (1st ed.). Piscataway: Wiley-IEEE Press.

[JITT16]   Jaber, M., Imran, M. A., Tafazolli, R., & Tukmanov, A. (2016). 5G backhaul challenges and emerging research directions: A survey. *IEEE Access, 4*, 1743–1766.

[JHA$^+$16]   Jalali, F., Hinton, K., Ayre, R., Alpcan, T., & Tucker, R. S. (2016). Fog computing may help to save energy in cloud computing. *IEEE Journal on Selected Areas in Communications, 34*(5), 1728–1739.

[Jan20]   Janßen, G. (2020). Quantum information theory lectures.

[JPC$^+$19]   Jiang, N., Pu, Y.-F., Chang, W., Li, C., Zhang, S., & Duan, L.-M. (2019). Experimental realization of 105-qubit random access quantum memory. *npj Quantum Information, 5*, 28.

[JL17]   Johansson, N., & Larsson, J. (2017). Efficient classical simulation of the Deutsch–Jozsa and Simon's algorithms. *Quantum Information Processing, 16*(9), 233.

[JBBB19]   Jones, T., Brown, A., Bush, I., & Benjamin, S. C. (2019). QuEST and high performance simulation of quantum computers. *Scientific Reports, 9*, 10736.

[JADW00]   Jozsa, R., Abrams, D. S., Dowling, J. P., & Williams, C. P. (2000). Quantum clock synchronization based on shared prior entanglement. *Physical Review Letters, 85*(9), 2010–2013.

[KMWY16]   Karumanchi, S., Mancini, S., Winter, A., & Yang, D. (2016). Quantum channel capacities with passive environment assistance. *IEEE Transactions on Information Theory, 62*(4), 1733–1747.

[KP17]   Kerenidis, I., & Prakash, A. (2017). Quantum recommendation systems. In C.H. Papadimitriou (Ed.), *8th Innovations in Theoretical Computer Science Conference (ITCS 2017). Leibniz International Proceedings in Informatics (LIPIcs)* (vol. 67, pp. 49:1–49:21). Wadern: Schloss Dagstuhl–Leibniz-Zentrum fuer Informatik.

[Ket19]   Kettimuthu, R. (2019). SeQUeNCe simulator of quantum network communication. https://doi.org/10.13140/RG.2.2.22583.55204.

[KC18]   Kim, T., & Choi, B.-S. (2018). Efficient decomposition methods for controlled R_n using a single ancillary qubit. *Scientific Reports, 8*(1), 1–7.

[Kim08]   Kimble, H. (2008). The quantum internet. *Nature, 453*, 1023–1030.

[Kit95]   Kitaev, A. Y. (1995). Quantum measurements and the Abelian stabilizer problem. arXiv:quantph/9511026.

[Kit97a]   Kitaev, A. Y. (1997). Quantum computations: Algorithms and error correction. *Uspekhi Matematicheskikh Nauk, 52*(6), 53–112.

[Kit97b]   Kitaev, A. Y. (1997). Quantum error correction with imperfect gates. In *Quantum Communication, Computing, and Measurement* (pp. 181–188). Berlin: Springer.

[Kit03]   Kitaev, A. Y. (2003). Fault-tolerant quantum computation by anyons. *Annals of Physics, 303*(1), 2–30.

[KSV02]   Kitaev, A. Y., Shen, A., Vyalyi, M. N. (2002). *Classical and quantum computation. Graduate Studies in Mathematics*. Providence: American Mathematical Soc.

[Kle07]   Klesse, R. (2007). Approximate quantum error correction, random codes, and quantum channel capacity. *Physical Review A, 75*(6), 062315.

[KMM13]   Kliuchnikov, V., Maslov, D., & Mosca, M. (2013). Fast and efficient exact synthesis of single qubit unitaries generated by Clifford and T gates. *Quantum Information & Computation., 13*(7–8), 607–630.

[KLV00]   Knill, E., Laflamme, R., & Viola, L. (2000). Theory of quantum error correction for general noise. *Physical Review Letters, 84*(11), 2525–2528.

[KKB⁺14]   Kómár, P., Kessler, E. M., Bishof, M., Jiang, L., Sørensen, A. S., Ye, J., et al. (2014). A quantum network of clocks. *Nature Physics, 10*(8), 582–587.

[KYP15]   Kubica, A., Yoshida, B., & Pastawski, F. (2015). Unfolding the color code. *New Journal of Physics, 17*(8), 083026.

[LvN⁺06]   Ladd, T. D., van Loock, P., Nemoto, W. J., Munro, K., & Yamamoto, Y. (2006). Hybrid quantum repeater based on dispersive CQED interactions between matter qubits and bright coherent light. *New Journal of Physics, 8*(9), 184–184 (2006).

[LALS20]   Leditzky, F., Alhejji, M. A., Levin, J., & Smith, G. (2020). Playing games with multiple access channels. *Nature Communications, 11*(1), 1497.

[LOW10]   Leung, D., Oppenheim, J., & Winter, A. (2010). Quantum network communication—the butterfly and beyond. *IEEE Transactions on Information Theory, 56*(7), 3478–3490 (2010)

[LSC⁺17]   Li, X., Samaka, M., Chan, H. A., Bhamare, D., Gupta, L., Guo, C., et al. (2017). Network slicing for 5G: Challenges and opportunities. *IEEE Internet Computing, 21*(5), 20–27.

[LJK⁺16]   Liu, J., Jiang, Z., Kato, N., Akashi, O., & Takahara, A. (2016). Reliability evaluation for NFV deployment of future mobile broadband networks. *IEEE Wireless Communications, 23*(3), 90–96.

[L̈99]   Löber, P. (1999). *Quantum channels and simultaneous ID coding.* Ph. D. Thesis, Universität Bielefeld, Fakultät für Mathematik, Bielefeld.

[LLY⁺19]   Lu, H., Li, Z., Yin, X., Zhang, R., Fang, X., Li, L., et al. (2019). Quantum network coding for general repeater networks. *npj Quantum Inf, 89*(5), 1–5.

[Luk03]   Lukin, M. D. (2003). Colloquium: Trapping and manipulating photon states in atomic ensembles. *Reviews of Modern Physics, 75*, 457–472.

[LST09]   Lvovsky, A. I., Sanders, B. C., & Tittel, W. (2009). Optical quantum memory. Nature Photonics, *Vol, 3*, 706714.

[MSNVM18]   Matsuo, T., Satoh, T., Nagayama, S., & Van Meter, R. (2018). Analysis of measurement-based quantum network coding over repeater networks under noisy conditions. *Physical Review A, 97*(6), 062328.

[Mic20]   Microsoft. (2020). Microsoft quantum development kit. https://www.microsoft.com/enus/quantum/development-kit.

[MSG⁺16]   Mijumbi, R., Serrat, J., Gorricho, J., Bouten, N., De Turck, F., & Boutaba, R. (2016). Network function virtualization: State-of-the-art and research challenges. *IEEE Communications Surveys Tutorials, 18*(1), 236–262.

[Mos15]   Mosonyi, M. (2015). Coding theorems for compound problems via quantum Rényi divergences. *IEEE Transactions on Information Theory, 61*(6), 2997–3012.

[MATN15]   Munro, W. J., Azuma, K., Tamaki, K., & Nemoto, K. (2015). Inside quantum repeaters. *IEEE Journal of Selected Topics in Quantum Electronics, 21*(3), 78–90.

[MLK⁺16]   Muralidharan, S., Li, L., Kim, J., Lütkenhaus, N., Lukin, M. D., & Jiang, L. (2016). Optimal architectures for long distance quantum communication. *Scientific Reports, 6*, 20463.

[QuT17]   NetSquid – the network simulator for quantum information using discrete events (2017). https://netsquid.org/.

[NC10]   Nielsen, M. A., & Chuang, I. L. (2010). *Quantum computation and quantum information* (10th ed.). Cambridge: Cambridge University Press.

[Nö19]   Nötzel, J. (2019). Entanglement-enabled communication. *IEEE Journal on Selected Areas in Information Theory, 1*(2), 401–415.

[HB14]   Nötzel, J., & Boche, H. (2014). Positivity, discontinuity, finite resources, nonzero error for arbitrarily varying quantum channels. In *IEEE International Symposium on Information Theory* (pp. 541–545).

[NMN+14]   Nunes, B. A. A., Mendonca, M., Nguyen, X., Obraczka, K., & Turletti, T. (2014). A survey of software-defined networking: Past, present, and future of programmable networks. *IEEE Communications Surveys Tutorials, 16*(3), 1617–1634.

[OBK+16]   O'Malley, P. J. J., Babbush, R., Kivlichan, I. D., Romero, J., McClean, J. R., Barends, R., et al. (2016). Scalable quantum simulation of molecular energies. *Phys. Rev. X, 6,* 031007.

[Orú14]   Orús, R. (2014). A practical introduction to tensor networks: Matrix product states and projected entangled pair states. *Annals of Physics, 349,* 117–158.

[Orú19]   Orús, R. (2019). Tensor networks for complex quantum systems. *Nature Reviews Physics, 1*(9), 538–550.

[PHS+18]   Pattaranantakul, M., He, R., Song, Q., Zhang, Z., & Meddahi, A. (2018). NFV security survey: From use case driven threat analysis to state-of-the-art countermeasures. *IEEE Communications Surveys Tutorials, 20*(4), 3330–3368.

[PS01]   Peres, A., & Scudo, P. F. (2001). Transmission of a cartesian frame by a quantum system. *Physical Review Letters, 87*(16), 167901.

[PV07]   Plenio, M. B., & Virmani, S. (2007). An introduction to entanglement measures. *Quantum Information and Computation, 7*(1), 1–51.

[HB20a]   Poor, H.V., Boche, H., & Schaefer, R. F. (2020). Denial-of-service attacks on communication systems: Detectability and jammer knowledge. *IEEE Transactions on Signal Processing, 68,* 3754–3768.

[Pre98a]   Preskill, J. (1998). Fault-tolerant quantum computation. In *Introduction to quantum computation and information* (pp. 213–269). Singapore: World Scientific.

[Pre98b]   Preskill, J. (1998). *Lecture notes for physics: Quantum information and computation* (p. 322). CreateSpace Independent Publishing Platform

[Pre12]   Preskill, J. (2012). Quantum computing and the entanglement frontier. Technical Report Caltech authors:20120516-084322874, California Institute of Technology.

[Pre18]   Preskill, J. (2018). Quantum computing in the NISQ era and beyond. *Quantum, 2,* 79.

[Pro20]   ProjectQ. (2020). *ProjectQ powerful open source software for quantum computing.,* 5. https://projectq.ch/.

[QS17]   Quek, Y., & Shor, P. W. (2017). Quantum and superquantum enhancements to two-sender, two-receiver channels. *Physical Review A, 95*(5), 052329. https://netsquid.org/.

[Ren84]   Rennels, D. A. (1984). Fault-tolerant computing — concepts and examples. *IEEE Transactions on Computers, C-33*(12), 1116–1129.

[RBSG16]   Richart, M., Baliosian, J., Serrat, J., & Gorricho, J. (2016). Resource slicing in virtual wireless networks: A survey. *IEEE Transactions on Network and Service Management, 13*(3), 462–476.

[RS16]   Ross, N. J. & Selinger, P. (2016). Optimal ancilla-free Clifford+T approximation of z-rotations. *Quantum Information & Computation, 16*(11–12), 901–953.

[RG02]   Rudolph, T., & Grover, L. (2002). A 2 rebit gate universal for quantum computing. In *arXiv:quant-ph/0210187.*

[RG03]   Rudolph, T., & Grover, L. (2003). Quantum communication complexity of establishing a shared reference frame. *Physical Review Letters, 91*(21), 217905.

[Rus02]   Ruskai, M. B. (2002). Inequalities for quantum entropy: A review with conditions for equality. *Journal of Mathematical Physics, 43*(9), 4358–4375.

[SAH+20]   Salek, F., Anshu, A., Hsieh, M.-H., Jain, R., & Fonollosa, J. R. (2020). One-shot capacity bounds on the simultaneous transmission of classical and quantum information. *IEEE Transactions on Information Theory, 66*(4), 2141–2164.

[SINVM16]   Satoh, T., Ishizaki, K., Nagayama, S., & Van Meter, R. (2016). Analysis of quantum network coding for realistic repeater networks. *Physical Review A, 93*(3), 032302.

[SB14a]   Schaefer, R. F., & Boche, H. (2014). Physical layer service integration in wireless networks: Signal processing challenges. *IEEE Signal Processing Magazine, 31*(3), 147–156.

[SB14b]   Schaefer, R. F., & Boche, H. (2014). Robust broadcasting of common and confidential messages over compound channels: Strong secrecy and decoding performance. *IEEE Transactions on Information Forensics and Security, 9*(10), 1720–1732.

[SW97]   Schumacher, B., & Westmoreland, M. D. (1997). Sending classical information via noisy quantum channels. *Physical Review A, 56*(1), 131.

[SNS16]   Scott-Hayward, S., Natarajan, S., & Sezer, S. (2016). A survey of security in software defined networks. *IEEE Communications Surveys Tutorials, 18*(1), 623–654.

[Sen12]   Sen, P. (2012). Achieving the Han-Kobayashi inner bound for the quantum interference channel by sequential decoding. In *2012 IEEE International Symposium on Information Theory Proceedings* (pp. 736–740).

[SMSC⁺15]   Shalm, L. K., Meyer-Scott, E., Christensen, B. G., Bierhorst, P., Wayne, M. A., Stevens, M. J., et al. (2015). Strong loophole-free test of local realism. *Physical Review Letters, 115*(25), 250402.

[SLPwL15]   Shang, T., Li, J., Pei, Z., & wei Liu, J. (2015). Quantum network coding for general repeater networks. *Quantum Information Processing, 14*, 3533–3552.

[Shi03]   Shi, Y. (2003). Both Toffoli and controlled-NOT need little help to do universal quantum computing. *Quantum Information and Computation, 3*(1), 84–92.

[Sho96]   Shor, P. W. (1996). Fault-tolerant quantum computation. In *Proceedings of 37th Conference on Foundations of Computer Science* (pp. 56–65).

[SAA⁺10]   Simon, C., Afzelius, M., Appel, J., Boyer de la Giroday, A., Dewhurst, S. J., Gisin, N., et al. (2010). Quantum memories. *The European Physical Journal D, 58*(1), 1–22.

[SR00]   Steane, A. M., & Rieffel, E. G. (2000). Beyond bits: The future of quantum information processing. *Computer, 33*(1), 38–45.

[Str69]   Strassen, V. (1969). Gaussian elimination is not optimal. *Numerische mathematik, 13*(4), 354–356.

[Tan19]   Tang, E. (2019). A quantum-inspired classical algorithm for recommendation systems. In *Proceedings of the 51st Annual ACM SIGACT Symposium on Theory of Computing, STOC 2019* (pp. 217–228). New York: Association for Computing Machinery.

[TAC⁺10]   Tittel, W., Afzelius, M., Chaneliére, T., Cone, R. L., Kröll, S., Moiseev, S. A., & Sellars, M. (2010). Photon-echo quantum memory in solid state systems. *Laser & Photonics Reviews, 4*(2), 244–267.

[Tsa11]   Tsang, M. (2011). Quantum nonlocality in weak-thermal-light interferometry. *Physical Review Letters, 107*, 270402.

[Val79]   Valiant, L. G. (1979). The complexity of computing the permanent. *Theoretical Computer Science, 8*(2), 189–201.

[vH58]   van Hove, L. (1958). Von neumann's contributions to quantum theory. *Bulletin of the American Mathematical Society, 64*, 95–99.

[vLAB⁺19]   van Loock, P., Alt, W., Becher, C., Benson, O., Boche, H., Deppe, C., et al. (2020). Extending quantum links: Modules for fiber- and memory-based quantum repeaters. *Advanced quantum technologies* (vol. 3, No. 11, p. 1900141). New York: Wiley.

[vLLS⁺06]   van Loock, P., Ladd, T. D., Sanaka, K., Yamaguchi, F., Nemoto, K., Munro, W. J., et al. (2006). Hybrid quantum repeater using bright coherent light. *Physical Review Letters, 96*, 240501.

[vLLMN08]   van Loock, P., Lütkenhaus, N., Munro, W. J., & Nemoto, K. (2008). Quantum repeaters using coherent-state communication. *Physical Review A, 78*, 062319.

[Van20]   Van Meter, R. (2020). QuISP quantum internet simulation package. https://aqua. sfc.wide.ad.jp/quispwebsite/.

[VLMN09]   Van Meter, R., Ladd, T. D., Munro, W. J., & Nemoto, K. (2009). System design for a long-line quantum repeater. *IEEE/ACM Transactions on Networking, 17*(3), 1002–1013.

[VNM07]   Van Meter, R., Nemoto, K., & Munro, W. (2007). Communication links for distributed quantum computation. *IEEE Transactions on Computers, 56*(12), 1643–1653.

[vNMR44]   von Neumann, J., Morgenstern, O., & Rubinstein, A. (1994). *Theory of Games and Economic Behavior (60th Anniversary Commemorative Edition).* Princeton: Princeton University Press.

[WLZ+19]   Wang, Y., Li, J., Zhang, S., Su, K., Zhou, Y., Liao, K. et al. (2019). Efficient quantum memory for single-photon polarization qubits. *Nature Photonics, 13*, 346–351.

[WS14]   Wecker, D., & Svore, K. M. (2014). LIQUi|>: A software design architecture and domain-specific language for quantum computing. arXiv:quant-ph/1402.4467.

[WEH18]   Wehner, S., Elkouss, D., & Hanson, R. (2018). Quantum internet: A vision for the road ahead. *Science, 362*(6412), eaam9288.

[Wen90]   Wen, X.-G. (1990). Topological orders in rigid states. *International Journal of Modern Physics B, 4*(02), 239–271.

[Wil13]   Wilde, M. M. (2013). Sequential decoding of a general classical-quantum channel. *Proceedings of the Royal Society A: Mathematical, Physical and Engineering Sciences, 469*(2157), 20130259.

[Wil17]   Wilde, M. W. (2017). *Quantum information theory* (2nd ed.). Cambridge: Cambridge University Press.

[WH11a]   Wilde, M. M., & Hsieh, M.-H. (2011). Public and private resource trade-offs for a quantum channel. *Quantum Information Processing, 11*(6), 1465–1501.

[WH11b]   Wilde, M. M., & Hsieh, M.-H. (2011). The quantum dynamic capacity formula of a quantum channel. *Quantum Information Processing, 11*(6), 1431–1463.

[Win78]   Winograd, S. (1978). On computing the discrete fourier transform. *Mathematics of Computation, 32*(141), 175–199.

[Win99]   Winter, A. (1999). Coding theorem and strong converse for quantum channels. *IEEE Transactions on Information Theory, 45*(7), 2481–2485.

[Win13]   Winter, A. (2013). Identification via quantum channels. In *Information theory, combinatorics, and search theory. Lecture notes in computer science* (pp. 217–233). Berlin: Springer.

[Win16]   Winter, A. (2016). Tight uniform continuity bounds for quantum entropies: Conditional entropy, relative entropy distance and energy constraints. *Communications in Mathematical Physics, 347*(1), 291–313.

[WBC15]   Wood, C. J., Biamonte, J. D., & Cory, D. G. (2015). Tensor networks and graphical calculus for open quantum systems. *Quantum Information and Computation, 15*(9–10), 759–811.

[WZ82]   Wootters, W. K., & Zurek, W. H. (1982). A single quantum cannot be cloned. *Nature, 299*(5886), 802–803.

[WCK+19]   Wu, X., Chung, J., Kolar, A., Wang, E., Zhong, T., Kettimuthu, R. et al. (2019). Simulations of photonic quantum networks for performance analysis and experiment design. In *2019 IEEE/ACM Workshop on Photonics-Optics Technology Oriented Networking, Information and Computing Systems (PHOTONICS)* (pp. 28–35).

[Wyn75]   Wyner, A. D. (1975). The wire-tap channel. *Bell System Technical Journal, 54*(8), 1355–1387.

[XGU⁺19]   Xiang, Z., Gabriel, F., Urbano, E., Nguyen, G. T., Reisslein, M., & Fitzek, F. H. P.
           (2019). Reducing latency in virtual machines: Enabling tactile internet for human-
           machine co-working. *IEEE Journal on Selected Areas in Communications, 37*(5),
           1098–1116.

[YCH18]   Yang, Y., Chiribella, G., & Hayashi, M. (2018). Quantum stopwatch: How to
           store time in a quantum memory. *Proceedings of the Royal Society of London
           A: Mathematical, Physical and Engineering Sciences, 474*(2213), 20170773.

[YCL⁺17]   Yin, J., Cao, Y., Li, Y.-H., Liao, S.-K., Zhang, L., Ren, J.-G., et al. (2017). Satellite-
           based entanglement distribution over 1200 kilometers. *Science, 356*(6343), 1140–
           1144.

Index

Printed in the United States
by Baker & Taylor Publisher Services